Lecture Notes in Economics and Mathematical Systems

560

Founding Editors:

M. Beckmann
H. P. Künzi

Managing Editors:

Prof. Dr. G. Fandel
Fachbereich Wirtschaftswissenschaften
Fernuniversität Hagen
Feithstr. 140/AVZ II, 58084 Hagen, Germany

Prof. Dr. W. Trockel
Institut für Mathematische Wirtschaftsforschung (IMW)
Universität Bielefeld
Universitätsstr. 25, 33615 Bielefeld, Germany

Editorial Board:

A. Basile, A. Drexl, H. Dawid, K. Inderfurth, W. Kürsten, U. Schittko

T0223596

Arndt von Schemde

Index and Stability in Bimatrix Games

A Geometric-Combinatorial Approach

 Springer

Author

Arndt von Schemde
Lilleborg gata 6
0480 Oslo
Norway
schemde@gmail.com

Library of Congress Control Number: 2005929878

ISSN 0075-8442
ISBN-10 3-540-26366-7 Springer Berlin Heidelberg New York
ISBN-13 978-3-540-26366-1 Springer Berlin Heidelberg New York

Springer is a part of Springer Science+Business Media

springeronline.com

© Springer-Verlag Berlin Heidelberg 2005
Printed in Germany

Typesetting: Camera ready by author
Cover design: *Erich Kirchner*, Heidelberg

Printed on acid-free paper   42/3130Di   5 4 3 2 1 0

To my parents

Preface

This work originates from my PhD thesis at the London School of Economics and Political Science. I am indebted to Bernhard von Stengel for his excellent supervision. He introduced me to the questions addressed in this work. I am thankful for the many hours of discussions, and also for his guidance and patience in times when results seemed far away.

Also, I am grateful to Srihari Govindan and Robert Wilson for useful advice and encouragement.

Furthermore, I would like to thank the members of the Mathematics Department for their general support. In particular, I thank Jackie Everid, David Scott and Mark Baltovic for their assistance, as well as Nic Georgiou and Luis Cereceda for their help on the final draft.

There were many people who supported me personally. Foremost, I would like to thank my parents and Ane S. Flaatten for always being there for me. I am also beholden to Philipp Beckmann for his advice, and to Philip Hochstrate for his inspiration.

Finally, I would like to thank the London School of Economics and Political Science (LSE), the Department of Mathematics at LSE and the UK Engineering and Physical Sciences Research Council (EPSRC) for financial support.

Oslo, July 2005   *Arndt von Schemde*

Contents

Introduction

Since Shapley (1974) introduced the index for equilibria, its importance in the context of game theory has been increasingly appreciated. For example, index theory can be a useful tool with regards to strategic characterisations of equilibria and equilibrium components. Demichelis and Ritzberger (2003) show that an equilibrium component can only be evolutionary stable if its index equals its Euler characteristic. At the same time, most of the existing literature on the index is technically demanding, and the amount of algebraic topology required is substantial. As a consequence, this literature is difficult to access for most economists and other applied game theorists.

The contribution of this thesis can be divided into two parts. The first part concerns methods and techniques. By introducing a new geometric-combinatorial construction for bimatrix games, this thesis gives a new, intuitive re-interpretation of the index. This re-interpretation is to a large extent self-contained and does not require a background in algebraic topology. The second part of this thesis concerns the relationship between the index and strategic properties. In this context, the thesis provides two new results, both of which are obtained by means of the new construction and are explained in further detail below. The first result shows that, in non-degenerate bimatrix games, the index can fully be described by a simple strategic property. It is shown that the index of an equilibrium is $+1$ if and only if one can add strategies with new payoffs to the game such that the equilibrium remains the unique equilibrium of the extended game. The second result shows that the index can be used to describe a stability property of equilibrium components.

For outside option components in bimatrix games, it is shown that such a component is hyperessential if and only if it has non-zero index.

The new geometric-combinatorial construction, which is referred to as the *dual construction*, can be described as follows. For an $m \times n$ bimatrix game, the construction translates the combinatorial structure of the best reply regions for both players into an $(m-1)$-simplex that is divided into simplices and labelled regions (see, for example, Figure 2.6 below). The simplices in the division account for the best reply structure of player II. The simplices themselves are divided into best reply regions for player I, accounting for the best reply structure of player I.

In this representation of bimatrix games, the Nash equilibria are represented by points that are completely labelled with all pure strategies of player I. Earlier constructions required the use of all pure strategies of both players as labels. The index is simply the local orientation of the labels around a completely labelled point (Figure 2.11). The Lemke-Howson algorithm, which builds the foundation for Shapley's original index definition, can be re-interpreted as a path-following algorithm in the new construction (Figure 2.8). Since the new construction is of dimension $m-1$, both the index and the Lemke-Howson algorithm can be visualised in dimension at most 3 for every $m \times n$ bimatrix game with $m \leq 4$.

But the construction does not merely yield an intuitive re-interpretation of the index and the Lemke-Howson algorithm. More significantly, it can disclose relationships between the index and strategic properties. In this context, this thesis provides, as mentioned, two new results.

As for the first result, it is shown that the index of an equilibrium is $+1$ if and only if it is the unique equilibrium of an extended game. The result proves a conjecture by Hofbauer (2000) in the context of equilibrium refinement. The proof is based on the idea that one can divide an $(m-1)$-simplex such that there exists only one completely labelled point which represents the index $+1$ equilibrium (Figure 4.7). Then such a division can be achieved as the dual construction of an extended game where strategies for player II are added (Figure 4.8).

The second result solves, for a special case, a problem that was open for some time. This problem addresses the question whether and how topological essentiality and game theoretic essentiality (Wu and Jiang (1962); Jiang (1963)) are related. Govindan and Wilson (1997b) argue that the resolution of this problem is highly relevant with respect to axiomatic studies: Imposing topological essentiality as an axiom in a decision-theoretic agenda is questionable if there is a gap between topological and strategic essentiality. Hauk and Hurkens (2002) construct a game with an outside option equilibrium component that has index zero but is essential. This demonstrates that topological essentiality is not equivalent to strategic essentiality. However, their example fails the requirement of hyperessentiality, i.e. the component is not essential in all equivalent games (Kohlberg and Mertens (1986)). The follow-up question is whether hyperessentiality is the game theoretic counterpart of topological essentiality. In this thesis, it is shown that this is the case for outside option equilibrium components in bimatrix games. That is, an outside option equilibrium component in a bimatrix game is hyperessential if and only if it has non-zero index. The proof is based on creating equivalent games by duplicating the outside option. An example presented in this thesis shows that one can create an outside option equilibrium component that has index zero and is essential in all equivalent games that do not contain duplicates of the outside option. However, it can be shown that the component fails the requirement of hyperessentiality if allowing duplicates of the outside option.

The proof of this result employs the combinatorial nature of the index for components of equilibria. In the framework of the dual construction, the index for components of equilibria is defined by a combinatorial division of a boundary into labelled best reply regions. This re-interpretation of the index for components is very similar to the index in the framework of the Index Lemma, a generalisation of Sperner's Lemma. For labellings as in the Index Lemma it is shown that, if the index of a boundary triangulation is zero, then there exists a labelled triangulation such that the triangulation does not contain a completely labelled simplex. The proof extends an index-zero boundary division of a polytope into labelled regions such that no point in the interior of the polytope is completely labelled. This extension is then translated into a triangulation (Figure 6.2). The proof for outside option components works similarly. Given an index-zero component, the dual of the component can be divided into labelled regions such that no point is completely labelled. It is then shown that such a division can be achieved as the dual construction of an equivalent game in which the outside option is duplicated and perturbed (Figure 6.10).

The concept of essentiality is strongly influenced by the theory of fixed points and essential fixed point components (Fort, 1950). In a parallel and independent work, Govindan and Wilson (2004) show that, for general N-player games and general equilibrium components, a component has non-zero index if and only if it is hyperessential. Their proof is based on a well-known result from fixed point theory that shows that a fixed point component is essential if and only if it has non-zero index (O'Neill, 1953). Their proof is technically very demanding. In contrast, the proof presented here for the special case provides a geometric intuition and does not require a knowledge of fixed point theory.

There is, however, a link between the combinatorial approach of this thesis and fixed point theory. This link is established via Sperner's Lemma (Sperner, 1928). The representation of bimatrix games in form of the dual construction reveals strong analogies with Sperner's Lemma. Sperner's Lemma is a classical result from combinatorial topology and is equivalent to Brouwer's fixed point theorem. Using the parallels of the dual construction with Sperner's Lemma it is shown that the existence of Nash equilibria in a non-degenerate bimatrix game is equivalent to Brouwer's fixed point theorem. On a similar topic, McLennan and Tourky (2004) derive Kakutani's fixed point theorem using the Lemke-Howson algorithm.

An additional result of this thesis, which does not involve the dual construction, is the construction of equilibrium components with arbitrary index. It is shown that for every integer q there exists a bimatrix game with an outside option equilibrium component that has index q. The construction is purely based on the properties of the index, and does not require knowledge of algebraic topology. This result originates from Govindan, von Schemde and von Stengel (2003).

The structure of this thesis is as follows. Chapter 1 introduces notations and conventions used throughout this work (Section 1.1). Sections 1.2 and 1.3 contain reviews of the Lemke-Howson algorithm and index theory. Section 1.4 shows how equilibrium components of arbitrary index can be constructed. Chapter 2 introduces the dual construction (Sections 2.1 and 2.2) and gives a re-interpretation of the index and the Lemke-Howson algorithm (Sections 2.3 and 2.4). Chapter 3 describes the parallels between the dual construction, Sperner's Lemma, and Brouwer's fixed point theorem. In Chapter 4, it is shown that the index for non-degenerate bimatrix games can be fully described by a strategic property. In Chapter 5, the dual construction is extended to outside option equilibrium components (Section 5.2). It also contains a review of the Index Lemma (Section 5.1). Finally, Chapter 6 investigates the relationship between the index and hyperessentiality. Section 6.1 considers index-zero labellings in the context of the Index Lemma. In Section 6.2, it is shown that an outside option equilibrium component is hyperessential if and only if it has non-zero index. A list of symbols is given at the end. Proofs and constructions are illustrated by figures throughout this work.

1

Equilibrium Components with Arbitrary Index

This chapter describes a method of constructing equilibrium components of arbitrary index by using outside options in bimatrix games. It is shown that for every integer q there exists a bimatrix game with an outside option equilibrium component that has index q. The construction is similar to the one used in Govindan, von Schemde and von Stengel (2003). That paper also shows that q-stable sets violate a symmetry property which the authors refer to as the *weak symmetry axiom*. The construction of equilibrium components of arbitrary index is the main result of this chapter.

The structure of this chapter is as follows. Section 1.1 introduces notational conventions and definitions that are used throughout this work. Section 1.2 gives a brief review of the classical Lemke-Howson algorithm that finds at least one equilibrium in a non-degenerate bimatrix game. Although the Lemke-Howson algorithm does not play a role in the construction of equilibrium components of arbitrary index, it can be used in the index theory for non-degenerate bimatrix games. Shapley (1974) shows that equilibria at the ends of a Lemke-Howson path have opposite indices. The Lemke-Howson algorithm also plays an important role in subsequent chapters when it is interpreted in a new geometric-combinatorial construction (see Chapters 2 and 3). Section 1.3 reviews the concept of index for Nash equilibria in both non-degenerate bimatrix games and general N-player games. Using basic properties of the index for components of Nash equilibria, Section 1.4 shows how equilibrium components of arbitrary index can be constructed as outside options in bimatrix games. It is shown that for every integer q there exists a bimatrix game with an equilibrium component that has index q (Proposition 1.6).

1.1 Preliminaries

The following notations and conventions are used throughout this work. The k-dimensional real space is denoted as $\mathbb{R}^k$, with vectors as column vectors. An $m \times n$ bimatrix game is represented by two $m \times n$ payoff matrices A and B, where the entries A_{ij} and B_{ij} denote the payoffs for player I and player II in the i-th row and j-th column of A and B. The set of pure strategies of player I is denoted by $I = \{1, \ldots, m\}$, and the set of pure strategies of player II is represented by $N = \{1, \ldots, n\}$. The rows of A and B are denoted a_i and b_i for $i \in I$, and the columns of A and B are denoted A_j and B_j for $j \in N$. The sets of mixed strategies for player I and player II are given by

$$X = \left\{ x \in \mathbb{R}^m \mid \mathbf{1}_m^\top x = 1,\ x_i \geq 0\ \forall\, i \in I \right\},$$
$$Y = \left\{ y \in \mathbb{R}^n \mid \mathbf{1}_n^\top y = 1,\ y_j \geq 0\ \forall\, j \in N \right\},$$

where $\mathbf{1}_k \in \mathbb{R}^k$ denotes the vector with entry 1 in every row. For easier distinction of the pure strategies, let $J = \{m+1, \ldots, m+n\}$, following Shapley (1974). Any $j \in N$ can be identified with $m + j \in J$ and vice versa. A *label* is any element in $I \cup J$. For notational convenience, the label j is sometimes used to refer to the pure strategy $j - m$ of player II if there is no risk of confusion.

X is a standard $(m-1)$-simplex that is given by the convex hull of the unit vectors $e_i \in \mathbb{R}^m$, $i \in I$, and Y is a standard $(n-1)$-simplex given by the convex hull of the unit vectors $e_{j-m} \in \mathbb{R}^n$, $j \in J$. The terms "$(m-1)$" and "$(n-1)$" refer to the dimension of the simplex. In general, an $(m-1)$-simplex is the convex hull of m affinely independent points in some Euclidian space. These points are the *vertices* of the simplex, and the simplex is said to be *spanned* by its vertices.

An *affine combination* of points $z_1, \ldots, z_m$ in an Euclidian space can be written as $\sum_{i=1}^{m} \lambda_i z_i$ with $\sum_{i=1}^{m} \lambda_i = 1$ and $\lambda_i \in \mathbb{R}$, $i = 1, \ldots, m$. A *convex combination* is an affine combination with the restriction $\lambda_i \geq 0$, $i = 1, \ldots, m$. A set of m points $z_1, \ldots, z_m$ is *affinely independent* if none of these points is an affine combination of the others. This is equivalent to saying that $\sum_{i=1}^{m} \lambda_i z_i = 0$

and $\sum_{i=1}^{m} \lambda_i = 0$ imply that $\lambda_1 = \ldots = \lambda_m = 0$. A convex set has *dimension* d if it has $d+1$, but no more, affinely independent points. A *k-face* of an $(m-1)$-simplex is the k-simplex spanned by any subset of $k+1$ vertices. The standard $(m-1)$-simplex spanned by the unit vectors in $\mathbb{R}^m$ is denoted by $\triangle^{m-1}$. So $X = \triangle^{m-1}$ and $Y = \triangle^{n-1}$.

For a mixed strategy $x \in X$, the support of x are the labels of those pure strategies that are played with positive probability in x. The support for $y \in Y$ is defined similarly. So

$$\text{supp}(x) = \{i \in I \mid x_i > 0\}, \quad \text{supp}(y) = \{j \in J \mid y_{j-m} > 0\}.$$

The strategy sets X and Y can be divided into best reply regions $X(j)$ and $Y(i)$. These are the regions in X where $j \in J$ is a best reply and the regions in Y where $i \in I$ is a best reply, so

$$X(j) = \left\{ x \in X \mid B_j^\top x \geq B_k^\top x \, \forall \, k \in J \right\}, \quad Y(i) = \{ y \in Y \mid a_i y \geq a_k y \, \forall \, k \in I \}.$$

The regions $X(j)$ and $Y(i)$ are (possibly empty) closed and convex regions that cover X and Y. For a point x in X the set $J(x)$ consists of the labels of those strategies of player II that are a best reply with respect to x. The set $I(y)$ is defined accordingly, so

$$J(x) = \{ j \in J \mid x \in X(j) \}, \quad I(y) = \{ i \in I \mid y \in Y(i) \}. \tag{1.1}$$

For $i \in I$, the set $X(i)$ denotes the $(m-2)$-face of X where the i-th coordinate equals zero. For $j \in J$, the set $Y(j)$ is defined as the $(n-2)$-face of Y where the $(j-m)$-th coordinate equals zero.

$$X(i) = \left\{ (x_1, \ldots, x_m)^\top \in X \mid x_i = 0 \right\}, Y(j) = \left\{ (y_1, \ldots, y_n)^\top \in Y \mid y_{j-m} = 0 \right\}.$$

Similar to (1.1), the sets $I(x)$ and $J(y)$ are defined as

$$I(x) = \{ i \in I \mid x \in X(i) \}, \quad J(y) = \{ j \in J \mid y \in Y(j) \}. \tag{1.2}$$

The labels $L(x)$ of a point $x \in X$ and the labels $L(y)$ of a point $y \in Y$ are defined as

$$L(x) = \{ k \in I \cup J \mid k \in X(k) \}, \quad L(y) = \{ k \in I \cup J \mid k \in Y(k) \}. \tag{1.3}$$

From (1.1) and (1.2) it follows that $L(x) = I(x) \cup J(x)$ and $L(y) = I(y) \cup J(y)$. So the labels of a point $x \in X$ are those pure strategies of player I that are played with zero probability in x and those strategies of player II that are best replies to x. Similarly, the labels of $y \in Y$ are those pure strategies of player II that are played with zero probability in y and those strategies of player I that are best replies to y.

Definition 1.1. *An $m \times n$ bimatrix game is called non-degenerate if for all $x \in X$ and $y \in Y$ the number of best reply strategies against x is at most the size of the support of x, and the number of best reply strategies against y is at most the size of the support of y, i.e. $|J(x)| \leq |\mathrm{supp}(x)|$ and $|I(y)| \leq |\mathrm{supp}(y)|$ for all $x \in X$ and $y \in Y$.*

It follows directly that in a non-degenerate game a point $x \in X$ can have at most m labels $L(x)$ and that a point y in Y can have at most n labels $L(y)$. Non-degeneracy implies that $X(j)$ and $Y(i)$ are either full-dimensional or empty (in which case a strategy is strictly dominated). For non-degenerate games the set of vertices $V \subset X$ is defined as those points in X that lie on some $(k-1)$-face of X and that have k pure best reply strategies in player II's strategy space. The set of vertices W in Y is defined accordingly, i.e.

$$V = \{v \in X \mid \mathrm{supp}(v) = k, \; |J(v)| = k\},$$
$$W = \{w \in Y \mid \mathrm{supp}(w) = k, \; |I(w)| = k\}.$$

Non-degeneracy implies that V is the set of those points in X that have exactly m labels, and W is the set of those points in Y that have exactly n labels. Notice that the unit vectors in $\mathbb{R}^m$ and $\mathbb{R}^n$, i.e. those representing the pure strategies in X and Y, are in V and W. An *edge* in X is defined by $m-1$ labels, and an edge in Y is defined by $n-1$ labels. For subsets $K, K' \subset I \cup J$ let

$$X(K) = \{x \in X \mid K \subset L(x)\}, \quad Y(K') = \{y \in Y \mid K' \subset L(y)\}. \tag{1.4}$$

That is, in case $|K| = m-1$ and $|K'| = n-1$, an edge in X is defined by $X(K)$, and an edge in Y is defined by $Y(K')$. If the game is non-degenerate, every edge in X and every edge in Y is a line segment.

The notion of vertices and edges comes from the study of polyhedra and polytopes (see e.g. Ziegler (1995)). In general, a *polyhedron H* is a subset of

$\mathbb{R}^d$ that is defined by a finite number of linear inequalities. If the dimension of H is d, then it is called full-dimensional. A polyhedron that is bounded is called a *polytope*. A *face* of a polytope P is the intersection of P with a hyperplane for which the polytope is contained in one of the two halfspaces determined by the hyperplane. If these faces are single points, they are called *vertices*, if they are 1-dimensional line segments, they are called *edges*. If the dimension of a face is one less than the dimension of the polytope, it is called *facet*.

For a bimatrix game with payoff matrix B for player II, one can define a polyhedron over player I's mixed strategy space X as follows.

$$H = \{(x,v) \in X \times \mathbb{R} \mid 1_m^\top x = 1, \ B^\top x \leq 1_n v, \ x_i \geq 0 \ \forall i \in I\} \qquad (1.5)$$

The polyhedron H is referred to as the *best reply polyhedron*. In a similar fashion, one can define the best reply polyhedron over Y using the payoff matrix A. Note that one can assume that all entries of A and B are strictly greater than zero, since adding a positive constant to the payoffs does not affect the Nash equilibria of a game. The polyhedron H is described by the upper envelope, that is, the maximum, of the expected payoffs for pure strategies of player II as functions of the mixed strategy played by player I.

Figure 1.1 depicts the polyhedron H for the payoff matrix

$$B = \begin{bmatrix} 6 & 4 & 1 \\ 1 & 3 & 5 \end{bmatrix}.$$

For example, the line that describes the facet with label 3 is given by the line between $v = 6$ for pure strategy 1, and payoff $v = 1$ for pure strategy 2. The labels of a point on the boundary of H are the "labels" of the linear inequalities that are binding in that point. A vertex of H is described by m binding linear inequalities, edges of H are described by $m - 1$ binding linear inequalities. Each $(m - 1)$-facet of the polyhedron H is defined by a single binding inequality and corresponds either to a best reply strategy of player II or to an unplayed strategy of player I. If H is projected onto X, it yields the division of X into best reply regions $X(j)$.

The above definitions can be illustrated using the 3×3 bimatrix game that is given by the following payoff matrices, taken from von Stengel (1999a).

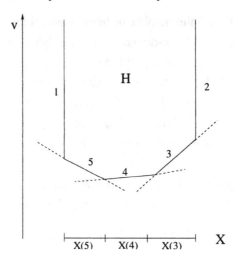

Fig. 1.1. The best reply polyhedron

$$A = \begin{bmatrix} 0 & 3 & 0 \\ 1 & 0 & 1 \\ -3 & 4 & 5 \end{bmatrix} \qquad B = \begin{bmatrix} 0 & 1 & -2 \\ 2 & 0 & 3 \\ 2 & 1 & 0 \end{bmatrix}. \tag{1.6}$$

The mixed strategy space X of player I is a 2-simplex, and so is the mixed strategy space Y of player II. Figure 1.2 shows the divisions of X and Y into best reply regions. For notational convenience, the subsets $X(k)$ and $Y(k)$, for $k \in I \cup J$, are just denoted by their label in Figure 1.2. The vertices $v \in V$ are emphasised by dots and are exactly those points in X that have three labels. A boundary 1-face of X carries the label of the pure strategy that is played with zero probability on that face. So, for example, the pure strategy $(0,0,1)^\top \in X$ has labels $\{1,2,4\}$, since strategies $1,2$ are played with zero probability, and strategy 4 of player II is the pure best reply strategy.

A *perturbation* of a bimatrix game is defined by two $m \times n$ matrices, ε_A and ε_B. The *perturbed game* is given by the game with payoff matrices $A + \varepsilon_A$ and $B + \varepsilon_B$. A perturbation is said to be small if $\|\varepsilon_A\|, \|\varepsilon_B\| \leq \varepsilon$ for some small $\varepsilon \geq 0$, where $\|\cdot\|$ denotes the Euclidian (or the maximum) norm on $\mathbb{R}^{mn}$. A perturbation is *generic* if the resulting perturbed game is non-degenerate.

The subsequent chapters use the concept of orientation as a definition of the index for Nash equilibria. For an m-tuple of vectors $\mathcal{V} = (v_1, \ldots, v_m)$ in $\mathbb{R}^m$, an orientation can be defined using the following term:

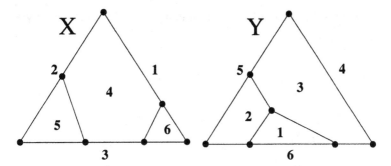

Fig. 1.2. The division of X and Y for the game in (1.6)

$$\text{sign det } \mathcal{V} = \text{sign det } \left[v_1 \ \ldots \ v_m \right]. \tag{1.7}$$

This term is $+1$ or -1 if and only if the vectors in $\mathcal{V}$ span an $(m-1)$-simplex that is contained in a hyperplane not containing $\mathbf{0} \in \mathbb{R}^m$. The two signs yield two equivalence classes of ordered vectors in general position. Choosing a standard orientation (which is usually that induced by the unit vectors $e_1, \ldots, e_m$), the orientation of $\mathcal{V}$ is $+1$ if it belongs to the same orientation class as the chosen standard orientation, and it is -1 otherwise.

The orientation can also be described as the sign of a permutation matrix. Suppose one has a set of m vectors that are in general position, and each vector has a distinct label $i \in \{1, \ldots, m\}$. Then the vectors can be ordered according to their labelling, and (1.7) can be applied to determine the orientation of the labelled set of vectors. Let the so-ordered set of vectors be denoted as $\mathcal{V}$. At the same time, one can re-order the vectors in such a way that (1.7) yields the same sign as that of the chosen standard orientation. Let this re-ordered basis be denoted as $\mathcal{V}'$. Both $\mathcal{V}$ and $\mathcal{V}'$ are a basis of $\mathbb{R}^m$, where one basis is a permutation of the other basis. The basis transformation is described by a permutation matrix D such that $\mathcal{V}' = D \cdot \mathcal{V}$, so det $\mathcal{V}' = \det D \cdot \det \mathcal{V}$. Hence det $D = +1$ if det $\mathcal{V}' = \det \mathcal{V}$, and det $D = -1$ if det $\mathcal{V}' = -\det \mathcal{V}$. So the determinant of the permutation matrix D, which is either $+1$ or -1, can also be used to describe the orientation. An illustration of the orientation concept is depicted in Figure 1.3. For the vectors v_1, v_2, v_3 as in Figure 1.3 the determinant has sign -1. The associated permutation of the labels, written as a product of cycles, is given by $(1)(23)$, and has also sign -1. This corresponds to an anti-clockwise orientation on $\triangle^2$ if looked at from the origin

$\mathbf{0} \in \mathbb{R}^3$, whereas the standard orientation induced by the unit vectors yields a clockwise orientation.

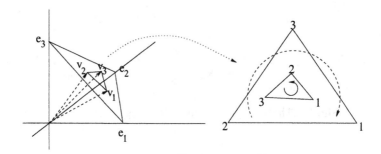

Fig. 1.3. The orientation of a basis

One can also define an orientation *relative* to a point $v_p \in \mathbb{R}^m$. Let $(v_1, \ldots, v_m)$ be an ordered m-tuple of vectors in $\mathbb{R}^m$. Then the orientation is defined by the term

$$\text{sign det } \mathcal{V} = \text{sign det } \left[v_1 - v_p \ldots v_m - v_p \right]. \tag{1.8}$$

Expression (1.7) is the same as (1.8) for $v_p = \mathbf{0} \in \mathbb{R}^m$. The term (1.8) is $+1$ or -1 if and only if the vectors in $v_1, \ldots, v_m, v_p$ span an m-simplex. That is, $v_1, \ldots, v_m$ span an $(m-1)$-simplex such that v_p is not an affine combination of the vectors $v_1, \ldots, v_m$. The hyperplane defined by the affine combinations of the vectors $v_1, \ldots, v_m$ divides $\mathbb{R}^m$ into two halfspaces. If two points v_p and v'_p lie in the same halfspace, the orientation relative to v_p and v'_p is the same. If the two points lie in different halfspaces, (1.8) yields opposite signs.

Let f be a function between two topological spaces S and T. If f is continuous then f is called a *mapping*. For two mappings f, g from a topological space S to a topological space T, i.e. $f, g : S \to T$, a *homotopy* h between f and g is a continuous deformation of f into g. A homotopy h can be described as a mapping $h : S \times [0, 1] \to T$ such that $h(x, 0) = f(x)$ and $h(x, 1) = g(x)$ for all $x \in S$. This is denoted as $f \simeq_h g$.

1.2 The Lemke-Howson Algorithm

In their seminal work, Lemke and Howson (1964) describe an algorithm for finding at least one equilibrium in a non-degenerate bimatrix game. This algorithm is referred to as the Lemke-Howson (L-H) algorithm, and it is the classical algorithm for finding Nash equilibria in non-degenerate bimatrix games. This section gives a brief review of the L-H algorithm, since it can be used in the theory of index for non-degenerate bimatrix games. Detailed reviews of the L-H algorithm can be found in Shapley (1974) and von Stengel (2002). Shapley (1974), motivated by the L-H algorithm, introduces the notion of index for non-degenerate bimatrix games. He shows that the equilibria at the two ends of an L-H path have opposite indices. The L-H algorithm also plays an important role in the subsequent chapters where it is translated into a new geometric-combinatorial construction (see Chapters 2 and 3).

Proposition 1.2. *Let G be an $m \times n$ bimatrix game (not necessarily non-degenerate). Then $(x,y) \in X \times Y$ is a Nash equilibrium of G if and only if $L(x) \cup L(y) = I \cup J$.*

Proof. This follows from the fact that in an equilibrium a pure strategy is a best reply strategy or is played with zero probability. If the game is degenerate, both might be the case. In any case, the condition $L(x) \cup L(y) = I \cup J$ ensures that only the best reply strategies are played with non-zero probability. $\qquad\square$

If a game is non-degenerate, an equilibrium strategy x plays a pure strategy with positive probability if and only if it is a best reply strategy against y, and vice versa. So in equilibrium $L(x) \cup L(y) = I \cup J$ and $L(x) \cap L(y) = \emptyset$. A pair (x,y) such that $L(x) \cup L(y) = I \cup J$ is called *completely labelled*.

The fact that an equilibrium strategy x plays a pure strategy with positive probability if and only if it is a best reply strategy against y (and vice versa) builds the basis for the L-H algorithm. The L-H algorithm describes a path in the product space $X \times Y$ along which the points are almost completely labelled with a fixed missing label. A pair (x,y) is said to be *almost completely labelled* if $L(x) \cup L(y) = I \cup J - \{k\}$ for some $k \in I \cup J$. The endpoints of a path are fully labelled and hence equilibria of the game. In order to obtain a starting point for the L-H algorithm one extends X and Y with the points

$\mathbf{0} \in \mathbb{R}^m$ and $\mathbf{0} \in \mathbb{R}^n$. These zero vectors can be seen as artificial strategies where the probability on each pure strategy is zero, i.e. no strategy is played. The pair $(\mathbf{0}, \mathbf{0})$ is then completely labelled.

The following description of the L-H algorithm follows that given by Shapley (1974). Let X_0 denote the boundary of the m-simplex spanned by $\mathbf{0} \in \mathbb{R}^m$ and $e_i \in \mathbb{R}^m$, $i \in I$. So X_0 consists of a union of $(m-1)$-faces, where one $(m-1)$-face of X_0 is given by X. The other $(m-1)$-faces of X_0 are spanned by vertices $\mathbf{0} \in \mathbb{R}^m$ and $e_i \in \mathbb{R}^m$, $i \in I - \{k\}$. Accordingly, the set Y_0 is defined as the boundary of the n-simplex spanned by $\mathbf{0} \in \mathbb{R}^n$ and $e_{j-m} \in \mathbb{R}^n$, $j \in J$. The $(n-1)$-face of Y_0 that is spanned by $e_{j-m} \in \mathbb{R}^n$, $j \in J$, represents Y. The other $(n-1)$-faces of Y_0 are spanned by vertices $\mathbf{0} \in \mathbb{R}^n$ and $e_{j-m} \in \mathbb{R}^m$, $j \in J - \{l\}$. For $x \in X_0$, the labels $L(x)$ are defined as $I(x) \cup J(x)$ for $x \in X$ and as $\{i \in I \mid x_i = 0\}$ otherwise. For $y \in Y_0$, the labels $L(y)$ are defined as $I(y) \cup J(y)$ for $y \in Y$ and as $\{j \in J \mid y_{j-m} = 0\}$ otherwise. The vertices in X_0 are the points with m labels, and the vertices in Y_0 are the points with n labels. So $\mathbf{0} \in \mathbb{R}^m$ is a vertex in X_0 with labels I and $\mathbf{0} \in \mathbb{R}^n$ is a vertex in Y_0 with labels J. The vertex pair $(\mathbf{0}, \mathbf{0}) \in \mathbb{R}^m \times \mathbb{R}^n$ is completely labelled, and it is referred to as the *artificial equilibrium*. For subsets $K, K' \subset I \cup J$, let

$$X_0(K) = \{x \in X_0 \mid K \subset L(x)\}, \; Y_0(K') = \{y \in Y_0 \mid K' \subset L(y)\}.$$

X_0 is a graph whose vertices are points with m labels, and whose edges are described by $m-1$ labels. Similarly, the set Y_0 is a graph whose vertices are points with n labels, and whose edges are described by $n-1$ labels. Depictions of X_0 and Y_0 for the game in (1.6) are given in Figure 1.4.

Now fix a label $k \in I \cup J$ and consider the subset of labels $I \cup J - \{k\}$. The idea of the L-H algorithm is to follow a unique path of almost completely labelled points with labels $I \cup J - \{k\}$ in the product graph $X_0 \times Y_0$. As a starting point, one chooses a completely labelled pair of vertices (x, y) in $X_0 \times Y_0$, so one can either start at an equilibrium or the artificial equilibrium. Each path with labels $I \cup J - \{k\}$ lies in the set

$$M(k) = \{(x, y) \in X_0 \times Y_0 \mid I \cup J - \{k\} \subset L(x) \cup L(y)\}. \qquad (1.9)$$

At the end of each path one finds another completely labelled pair of vertices, i.e. an equilibrium. The paths of almost completely labelled points are

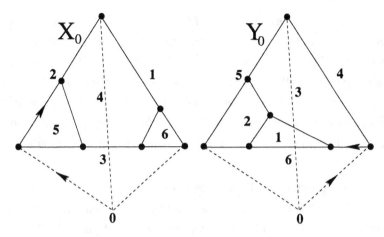

Fig. 1.4. The L-H algorithm for the game in (1.6)

referred to as *L-H paths*. The following theorem and proof can also be found in von Stengel (2002).

Theorem 1.3 (Lemke and Howson, 1964; Shapley, 1974). *Let G be a non-degenerate bimatrix game and k be a label in $I \cup J$. Then $M(k)$ as in (1.9) consists of disjoint paths and cycles in the product graph $X_0 \times Y_0$. The endpoints of the paths are the equilibria of the game and the artificial equilibrium $(\mathbf{0},\mathbf{0})$. The number of equilibria is odd.*

Proof. Let $(x,y) \in M(k)$. Then x and y have together either $m+n$ or $m+n-1$ labels. In the former case, the tuple (x,y) is either an equilibrium or the artificial equilibrium. In the latter case, one has $L(x) \cup L(y) = I \cup J - \{k\}$, and there are the following three possibilities:

a) $|L(x)| = m$ and y has $n-1$ labels. Then x is a vertex in X_0, and y lies on some edge $e(y)$ in Y_0. So $\{x\} \times e(y)$ is an edge in $X_0 \times Y_0$.
b) x has $m-1$ labels and is part of an edge $e(x)$ in X_0, while y has n labels and is a vertex in Y_0. Then $e(x) \times \{y\}$ is an edge in $X_0 \times Y_0$.
c) x has m labels and y has n labels. So (x,y) is a vertex in the product graph $X_0 \times Y_0$.

Therefore, the set $M(k)$ defines a subgraph of $X_0 \times Y_0$. If (x,y) is completely labelled, then the vertex (x,y) is incident to a unique edge in the subgraph $M(k)$, namely $\{x\} \times Y_0(L(y) - \{k\})$ if $k \in L(y)$ or $X_0(L(x) - \{k\}) \times \{y\}$ if

$k \in L(x)$. In case c), one has $L(x) \cup L(y) = I \cup J - \{k\}$, so there must be a duplicate label in $L(x) \cap L(y)$. But this means that (x,y) is incident to both edges $\{x\} \times Y_0(L(y) - \{k\})$ and $X_0(L(x) - \{k\}) \times \{y\}$. Therefore, the set $M(k)$ is a subgraph where all vertices are incident to one or two edges. Hence, the subgraph $M(k)$ consists of paths and cycles. The endpoints of the paths are the equilibria and the artificial equilibrium. Since the number of the endpoints is even, the number of equilibria is odd (not counting the artificial equilibrium). □

The L-H algorithm can be illustrated by the game in (1.6). This is depicted in Figure 1.4. One starts in the completely labelled artificial equilibrium $(\mathbf{0}, \mathbf{0})$. Now choose a label to drop, say label 1 of player I. This determines an edge in X_0 along which the points have labels $2,3$. At the other end of this edge one finds a vertex $v \in X_0$ with labels $2,3,5$. The vertex pair $(v, \mathbf{0})$ has labels $2,3,5$ and $4,5,6$, so 5 is a duplicate label. This determines an edge in Y_0 with labels $4,6$ leading to the vertex w with labels $3,4,6$. So the vertex pair (v,w) has the duplicate label 3, and one follows the edge in X_0 that is given by labels $2,5$, leading to v' with labels $2,4,5$. Now (v',w) has duplicate label 4. This yields an edge in Y_0 defined by labels $6,3$, leading to w' with labels $6,1,3$. The pair (v',w') is completely labelled and hence an equilibrium of the game in (1.6).

1.3 Index Theory

For non-degenerate bimatrix games, the index for equilibria was first introduced by Shapley (1974). Shapley's index theory is motivated by the L-H algorithm, and Shapley shows that equilibria which are connected via an L-H path have opposite indices.

Formally, let (x,y) be an equilibrium of a non-degenerate bimatrix game with payoff matrices A and B. Let A' and B' denote the square sub-matrices obtained from A and B by deleting those rows and columns that correspond to pure strategies played with zero probability in x and y. So

$$A' = [A_{ij}]_{i \in \text{supp}(x) \wedge j \in \text{supp}(y)}, \qquad B' = [B_{ij}]_{i \in \text{supp}(x) \wedge j \in \text{supp}(y)} \qquad (1.10)$$

are the payoff matrices restricted to the support of x and y. Without loss of generality it can be assumed that all entries of A and B are (strictly) greater than zero. This is possible since adding a positive constant to the entries of A or B does not affect the equilibria of the game.

Definition 1.4 (Shapley, 1974). *The index of an equilibrium* (x,y) *of a non-degenerate bimatrix game with payoff matrices A and B is given as the negative of the sign of the determinant of the following index matrix obtained from A and B:*

$$I(x,y) = -\text{sign det} \begin{bmatrix} 0 & B' \\ (A')^\top & 0 \end{bmatrix}.$$

Using basic laws for the calculation of the determinant, this expression simplifies to $I(x,y) = \text{sign}(-1)^{k+1}\det(A')^\top \det B'$, where k is the size of the support of x and y.

Remark 1.5. Shapley (1974) defines the index as

$$\text{sign det} \begin{bmatrix} 0 & B' \\ (A')^\top & 0 \end{bmatrix},$$

i.e. Definition 1.4 is the negative of the original definition, for the following reasons. Definition 1.4 is consistent with the generalisation of the index for components of equilibria. Furthermore, according to Definition 1.4, pure strategy equilibria and equilibria that are the unique equilibrium of a game have index $+1$.

Shapley shows that equilibria that are connected via an L-H path have opposite indices and that the sum of indices of equilibria of a game equals $+1$ (using the index as in Definition 1.4). In Shapley's original work, the proof of this claim is not very intuitive. A more intuitive approach can be found in Savani and von Stengel (2004). Basically, it employs the fact that along a path with $m+n-1$ labels that connects two completely labelled vertices the "relative position" of the labels stays constant. This is illustrated in Figure 1.5. The two fully labelled points are connected via a path with labels $2,3$, where 2 is always on the left of the path and 3 on the right (and the non-missing labels have a similar fixed orientation in higher dimension). The fully labelled vertex on the left reads $1,2,3$ in clockwise orientation, and the fully labelled vertex on the right reads $1,2,3$ in anti-clockwise orientation. In this sense the index is an orientation of the labels around a fully labelled vertex.

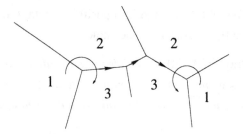

Fig. 1.5. Equilibria at the ends of L-H paths have opposite indices

To apply this concept of orientation to bimatrix games, Savani and von Stengel first consider symmetric games. In symmetric games, the L-H paths can be followed in the strategy space of just one player, say player I, by replacing the labels of player II in X by the corresponding best reply labels of player I in the division of Y. Then the Nash equilibria of a symmetric game correspond to vertices in X that have labels $1, \cdots, m$. For the 3×3 coordination game, this is depicted in Figure 1.6. But every non-symmetric game with payoff matrices A and B can be symmetrised by constructing the game with payoff matrices

$$C = \begin{bmatrix} 0 & A \\ B^\top & 0 \end{bmatrix}, \qquad C^\top = \begin{bmatrix} 0 & B \\ A^\top & 0 \end{bmatrix},$$

again assuming that all payoffs of A and B are strictly greater than 0. Then the equilibria of the game with matrices C and $C^\top$ correspond to the equilibria of the original game by restricting the solutions of the symmetrised game to X and Y, and re-normalising the probabilities.

In non-degenerate games, the Nash equilibria are singletons in the product space $X \times Y$. For degenerate games one has to consider sets of equilibria in $X \times Y$. Kohlberg and Mertens (1986, Proposition 1) show that the set of Nash equilibria of any finite game has finitely many connected components. A maximally connected set of Nash equilibria is referred to as a *component of equilibria*. The index of a component of equilibria of a game is an integer that is computed as the local degree of a map for which the Nash equilibria of the game are the zeros. Loosely speaking, the local degree of a map counts the number of cycles (in higher dimension spheres) around zero obtained by the image of a cycle (in higher dimension sphere) around the component (see

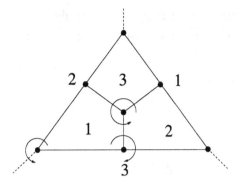

Fig. 1.6. The index in the coordination game

e.g. Dold (1972, IV, 4)). The Nash equilibria of a game can be described as the fixed points of a mapping $f : X \times Y \to X \times Y$ (see e.g. Nash (1951) or Gül, Pearce and Stacchetti (1993) for such mappings). Such maps are called *Nash maps*. Defining $F = f - Id$ yields a *Nash field* whose zeros are the Nash equilibria of a game. The index is independent of the particular map used (see Govindan and Wilson (1997b), for bimatrix games, and, for games with any number of players, Demichelis and Germano (2000)). For generic bimatrix games it is the same as the index in Definition 1.4 (Govindan and Wilson (1997b)). An introduction to the concept of index for components of equilibria can be found in Ritzberger (2002, 6.5).

Using the Kohlberg-Mertens (K-M) structure theorem (Kohlberg and Mertens (1986, Theorem 1)), the index can also be expressed as the local degree of the projection map from the equilibrium correspondence to the space of games (see Govindan and Wilson (1997a), for bimatrix games, and, for games with any number of players, Demichelis and Germano (2000)). This can be illustrated using the following parameterised game.

$$G(t) = \begin{bmatrix} 1-t, 1-t & 0,0 \\ 0,0 & t,t \end{bmatrix} \tag{1.11}$$

In this example, the games $G(t)$ are parameterised by $t \in \mathbb{R}$. Figure 1.7 shows that the equilibrium correspondence $E(G(\cdot)) \subset G(\cdot) \times (X \times Y)$ over $G(\cdot)$ is homeomorphic to $G(\cdot)$ itself. In Figure 1.7, p denotes the probability for the first strategy of player I in equilibrium. If player I plays $(p, 1-p) \in X$ in an equilibrium, then player II's strategy in that equilibrium is also $(p, 1-p) \in Y$, where $p = t$ gives the mixed equilibrium of the game when $0 < t < 1$.

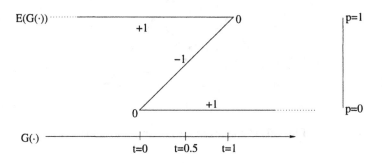

Fig. 1.7. The K-M structure theorem

In general, let Γ denote the space of games for a fixed number of players with a fixed number of strategies. Then Γ can be parameterised by $\mathbb{R}^k$, where k equals the number of players multiplied by the product of the numbers of pure strategies per player. Let Σ denote the product space of mixed strategy spaces. Then the equilibrium correspondence over Γ is defined as

$$E(\Gamma) = \{(G, \sigma) \in \Gamma \times \Sigma \mid \sigma \text{ is an equilibrium of } G\}.$$

The K-M structure theorem states that the space of games Γ is homeomorphic to $E(\Gamma)$ (after a one-point compactification). In general, the K-M structure theorem does not apply to restrictions of the space of games Γ as in (1.11). If, for example, one restricts Γ to a single point that represents a game with more than one component of equilibria, the space of games, i.e. the single point, is not homeomorphic to the graph of the equilibrium correspondence, which consists of several disjoint sets of equilibria. Nevertheless, (1.11) gives a good illustration of the K-M structure theorem.

For the illustration in Figure 1.7, the local degree of the projection map from $E(\Gamma)$ on Γ measures, loosely speaking, the local orientation of the equilibrium correspondence relative to the orientation of Γ. In the example, all completely mixed equilibria have index -1. The pure equilibria in the non-degenerate games (i.e. $t \notin \{0, 1\}$) have index $+1$. The corners of the Z-shaped correspondence are those pure strategy equilibria in the degenerate games

($t \in \{0, 1\}$) which disappear or split into two equilibria with opposite indices for small perturbations. These have index 0.

The index for components and for singletons in the non-degenerate case has useful properties that are employed in the next section to construct components of arbitrary index.

1) For the non-degenerate case, the index defined as the local degree is the same as the index defined in Definition 1.4 (Govindan and Wilson (1997b)).

2) The sum of indices of components of equilibria for a fixed game equals $+1$ (see e.g. Govindan and Wilson (1997a)).

3) For sufficiently small generic perturbations of a degenerate game, the index of a component equals the sum of indices of equilibria in the perturbed game close to the component (see e.g. Govindan and Wilson (1997a;b) for a discussion). This fact is illustrated in Figure 1.7. Take the pure strategy equilibrium in the degenerate case $t = 1$ that has index 0. If the game is perturbed "to the right" ($t + \varepsilon$) the equilibrium vanishes, if it is perturbed "to the left" ($t - \varepsilon$) it splits into two equilibria close to it, one with index -1 and one with index $+1$.

4) The index of a component is the same in all equivalent games (Govindan and Wilson (1997a, Theorem 2; 2004, Theorem A.3)), i.e. it is invariant under adding convex combinations of existing strategies with the respective payoffs as new pure strategies.

An equilibrium component is said to be *essential* if every small perturbation of the game yields a perturbed game that has equilibria close to the component. It follows that an equilibrium component with non-zero index is essential. An equilibrium component is said to be *hyperessential* if it is essential in all equivalent games. Therefore an equilibrium component with non-zero index is also hyperessential. Chapter 6 reviews the concept of (hyper)essentiality in more detail. It addresses the question whether and under what circumstances the converse is also true, i.e. whether (hyper)essentiality implies non-zero index.

1.4 Construction of Equilibrium Components with Arbitrary Index

In this section it is shown how games with equilibrium components of arbitrary index can be constructed. This new result is based on a construction that uses outside options in bimatrix games. The construction is similar to the one used in Govindan, von Schemde and von Stengel (2003), where the authors construct symmetric components of arbitrary index in order to show that q-stability violates a notion of symmetry. A great part of the following description is borrowed from this paper.

First, consider a 2×2 coordination game, say

$$H^2 = \begin{bmatrix} 10,10 & 0,0 \\ 0,0 & 10,10 \end{bmatrix}$$

(in agreement with the notation in (1.16) below). This game has two pure strategy equilibria, and one mixed equilibrium, where both players play the mixed strategy $(\frac{1}{2}, \frac{1}{2})$. The index of any of these equilibria is easily determined by the following two properties, which hold for any game: A pure strategy equilibrium which is *strict* (that is, all unplayed pure strategies have a payoff that is strictly lower than the equilibrium payoff) has index $+1$; The sum over all equilibria of their indices is $+1$. Therefore, the mixed equilibrium in H^2 has index -1. This can also be verified using Definition 1.4.

Next, an *outside option* called *Out* is added to the set of pure strategies of player II, say, giving the game

$$G^- = \begin{bmatrix} 10,10 & 0,0 & 0,9 \\ 0,0 & 10,10 & 0,9 \end{bmatrix}. \tag{1.12}$$

An outside option can be thought of as an initial move that a player can make which terminates further play, and gives a constant payoff to both players. If the player has not chosen his outside option, the original game is played. The outside option payoff above is 9 for player II. This has the effect that an equilibrium of the original game with payoff less than 9 for player II disappears, in this case the mixed strategy equilibrium. Geometrically, one can consider the upper envelope, i.e. the maximum of the expected payoffs for the pure strategies of player II, as functions of the mixed strategy played by player I as described in Section 1.1. Any equilibrium strategy of player I, together with its payoff to player II, is on that upper envelope. The outside option gives an additional constant function that "cuts off" any former equilibrium payoffs below it. This is depicted in Figure 1.8. It shows the upper envelope of the expected payoffs for pure strategies of player II and the resulting division of player I's strategy space X before and after adding Out to player II's strategy space.

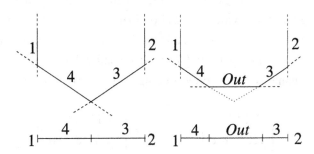

Fig. 1.8. Division of X before and after adding an outside option

In game G^-, the original pure strategy equilibria of H^2 are unaffected, and continue to have index $+1$. Any such equilibrium, as long as it remains (quasi-)strict after introducing the outside option, keeps its index, as the index of a strict equilibrium can be defined in terms of the payoff sub-matrices corresponding to the pure best replies (see Definition 1.4). The mixed strategy equilibrium of H^2 is absorbed into an equilibrium *component* where player II plays his last strategy Out. The original mixed equilibrium strategy $(\frac{1}{2}, \frac{1}{2})$ of player I is part of the outside option component, which is given by the set of mixed strategies of player I so that Out is a best response. In G^- above, it is easy to see that these are all mixed strategies of player I where each pure strategy has probability at most $9/10$. In general, the outside option component is defined by a set of linear inequalities, one for each pure strategy of the player who plays Out.

Let G be some game with an outside option. Then the outside option equilibrium component of the game G by is denoted by $C(G)$. In (1.12), the index of $C(G^-)$ is -1, which is simply the sum of the indices of all equilibria of the original game H^2 that have been absorbed into the outside option component, because the sum of all indices is $+1$. As described in Section 1.3, the index of an equilibrium component also equals the sum of indices of equilibria near the component when payoffs are perturbed generically; this sum does not depend on the perturbation.

It is well-known that the best response structure of a bimatrix game remains unchanged when adding a constant to any column of the payoffs to the row player, or a constant to a row of the column player's payoffs. This will allow to cut off pure strategy equilibria rather than mixed equilibria by using an outside option. Start with a 2×2 coordination game with payoffs $1, 1$ on and $0, 0$ off the main diagonal, and add the constant 12 to the first column of player I and row of player II, and 7 to the second column respectively row. The resulting game H and a corresponding outside option game G are given by

$$H = \begin{bmatrix} 13,13 & 7,12 \\ 12,7 & 8,8 \end{bmatrix}, \qquad G = \begin{bmatrix} 13,13 & 7,12 & 0,9 \\ 12,7 & 8,8 & 0,9 \end{bmatrix}.$$

The game H has two pure equilibria with payoffs $13, 13$ and $8, 8$, respectively, and one mixed equilibrium where both play $(\frac{1}{2}, \frac{1}{2})$ with payoffs $10, 10$. The outside option with payoff 9 for player II cuts off the pure strategy equilibrium with payoffs $8, 8$ but leaves the other equilibria intact. Consequently, the component $C(G)$ has index $+1$.

Next, one can "destroy" the pure strategy equilibrium in G by adding another row to the game. Consider the games

$$H' = \begin{bmatrix} 13,13 & 7,12 \\ 12,7 & 8,8 \\ 14,1 & 1,2 \end{bmatrix}, \qquad G' = \begin{bmatrix} 13,13 & 7,12 & 0,9 \\ 12,7 & 8,8 & 0,9 \\ 14,1 & 1,2 & 0,9 \end{bmatrix}.$$

Compared to H, the pure strategy equilibrium with payoffs $13, 13$ is no longer present in H'. It is replaced by another, mixed equilibrium where player II plays $(\frac{6}{7}, \frac{1}{7})$ and player I plays $(\frac{1}{2}, 0, \frac{1}{2})$, with payoffs 7 to player II and $85/7$ to player I. This new mixed equilibrium has index $+1$. Since the payoff to player II in that equilibrium is less than the outside option payoff 9, that equilibrium disappears in G'. Consequently, the component $C(G')$ has index $+2$, because the only equilibrium that is not cut off has index -1.

Finally, consider the following game H^-, which is a symmetrised version of H':

$$H^- = \begin{bmatrix} 13,13 & 7,12 & 1,14 \\ 12,7 & 8,8 & 2,1 \\ 14,1 & 1,2 & 1,1 \end{bmatrix}. \tag{1.13}$$

In this game, the mixed strategy equilibrium where both players play $(\frac{1}{2},\frac{1}{2},0)$ is the equilibrium with the highest payoff, yielding 10 for both players. This equilibrium has index -1. The other equilibria are as follows: The mixed strategy $(\frac{1}{2},0,\frac{1}{2})$ of player I, which together with $(\frac{6}{7},\frac{1}{7})$ of player I forms an equilibrium of H', is no longer part of an equilibrium as the third strategy of player II in H^- gives a higher payoff. By playing that strategy as well, one obtains a *completely mixed* equilibrium where both players play $(\frac{1}{2},\frac{1}{12},\frac{5}{12})$, with resulting payoff $15/2$ to both players. This equilibrium has index $+1$, as has the pure strategy equilibrium with payoffs $8,8$. There are no other equilibria of H^-.

H^- is used for constructing components with arbitrarily high positive index. For $k \geqslant 1$, let H^{-k} be the game consisting of k copies of the game H^- on the diagonal and zeros everywhere else, that is,

$$H^{-k} = \underbrace{\begin{bmatrix} H^- & 0,0 & \cdots & 0,0 \\ 0,0 & H^- & & 0,0 \\ \vdots & & \ddots & \vdots \\ 0,0 & 0,0 & \cdots & H^- \end{bmatrix}}_{k \text{ copies}}. \tag{1.14}$$

Each player has $3k$ strategies in H^{-k}. For any nonempty set of the k copies of H^-, and any equilibrium in such a copy, one obtains an additional equilibrium of H^{-k} by suitable probability weights assigned to the copies. All such mixtures involving more than one copy, however, give payoffs less than 8. There are no other equilibria of H^{-k} as the payoffs in a copy of H^- are all positive, and the other payoffs are zero.

The superscript in H^{-k} indicates the sum of indices of those equilibria that are not cut off by adding a suitable outside option. The outside option is, as before, added to player II's strategy space, and is also referred to as *Out* as an additional pure strategy. This gives the game

$$G^{k+1} = \begin{bmatrix} & & 0,9 \\ H^{-k} & & \vdots \\ & & 0,9 \end{bmatrix}. \tag{1.15}$$

The game G^{k+1} has $k+1$ equilibrium components: the k mixed strategy equilibria where both players play strategies 1 and 2 in one copy of H^- with probability $\frac{1}{2}$ (yielding a payoff of 10 for both), and the equilibrium component in which player II chooses the last strategy, the outside option *Out*. That component $C(G^{k+1})$ is given by those strategy pairs where player II plays *Out*, and player I playing such that *Out* is a best response. All isolated equilibria have index -1. Since the indices of all equilibrium components have to add up to one, the outside option equilibrium component $C(G^{k+1})$ has index $k+1$, which is chosen as a superscript for G in (1.15). Therefore, for each positive integer q, the game G^q in (1.15) has a component with index q; this includes the trivial case $q=1$ and $k=0$, which is a 1×1 game.

The division of player I's mixed strategy space X for the game G^2 is depicted in Figure 1.9. It shows that, except for the equilibrium vertex $(\frac{1}{2}, \frac{1}{2}, 0) \in X$, all other vertices that are part of an equilibrium in H^- are cut off by the outside option.

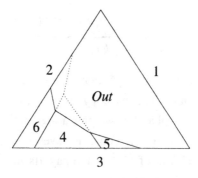

Fig. 1.9. The division of X for the game G^2 with outside option

A similar, simpler construction gives equilibrium components with arbitrary negative index. For $k \geqslant 2$, let H^k be the following $k \times k$ game:

$$H^k = \begin{bmatrix} 10,10 & 0,0 & \cdots & 0,0 \\ 0,0 & 10,10 & & 0,0 \\ \vdots & & \ddots & \vdots \\ 0,0 & 0,0 & \cdots & 10,10 \end{bmatrix} \tag{1.16}$$

$$\underbrace{}_{k \text{ columns}}$$

Just as (1.15) is obtained from (1.14), one can add an outside option for player II, and obtain

$$G^{-(k-1)} = \begin{bmatrix} & 0,9 \\ H^k & \vdots \\ & 0,9 \end{bmatrix} \qquad (k \geqslant 2). \tag{1.17}$$

The equilibria of game $G^{-(k-1)}$ are the k pure strategy equilibria of the coordination game, yielding a payoff of 10 for both players, and the outside option equilibrium component $C(G^{-(k-1)})$ (see Figure 1.8 for the case $k = 2$). Since pure strategy equilibria have index $+1$, it follows that $C(G^{-(k-1)})$ has index $-(k-1)$.

Hence, for each negative integer q, there exists a game that has an equilibrium component with index q. The case $k = 1$ gives an empty equilibrium component (which can be thought of as having index 0), since in this case the first strategy by player II strictly dominates Out. Therefore it is required that $k \geqslant 2$ in (1.17).

From the above, one can now easily construct a game with a non-trivial equilibrium component that has index 0. This is done by combining the games H^k and $H^{-(k-1)}$ in a new game by placing them on the diagonal, and adding an outside option for player II as before. The case $k = 2$ is sufficient, so let G^0 be the following 5×6 game:

$$G^0 = \begin{bmatrix} H^2 & 0 & 0,9 \\ 0 & H^- & 0,9 \end{bmatrix}. \tag{1.18}$$

As argued after (1.14), the only equilibria in G^0 that are not cut off are those with payoffs $10, 10$ in H^2 or H^-. Thus, by a counting argument, the outside option equilibrium component $C(G^0)$ has index 0. The constructions prove the following proposition.

Proposition 1.6. *For each integer q, there exists a (bimatrix) game that has a component of equilibria with index q.*

In general, index 0 components are easy to construct (see also $k = 1$ in (1.17) for the trivial case). Consider for example the game

$$\begin{bmatrix} 1,1 \; 0,0 \\ 0,0 \; 0,0 \end{bmatrix}.$$

This game is the same as $G(0)$ in (1.11) and has two pure strategy equilibria, one with payoff 1 and the other one with payoff 0. It is easy to verify that the equilibrium with payoff 1 has index $+1$. It "survives" every small payoff perturbation. The pure strategy equilibrium with payoff 0 has index zero. The payoffs can be perturbed such that this equilibrium either vanishes or splits into two equilibria with opposite indices (see also Figure 1.7). The reason for providing G^0 as in (1.18) is that a similar construction is used in Govindan et al. (2003) in order to show that 0-stable sets violate a notion of symmetry. Furthermore, in Chapter 6 it is shown that the outside option equilibrium component of the game G^0 is essential in all equivalent games that do not contain a duplicate of *Out*. However, it is not hyperessential when allowing copies of *Out*.

2

A Reformulation of the Index for Equilibria in Bimatrix Games

This chapter introduces a new geometric-combinatorial construction for non-degenerate bimatrix games that allows one to give a new characterisation of Nash equilibria and index in bimatrix games. Given an $m \times n$ non-degenerate bimatrix game (assuming $m \leq n$ without loss of generality), the construction yields a division of an $(m - 1)$-simplex in which the Nash equilibria and the index can be characterised by the labels of player I only. So, for example, any $3 \times n$ bimatrix game can be represented by a division of a 2-dimensional simplex using only labels $1, 2, 3$.

The new construction, which is referred to as the *dual construction*, allows an intuitive definition of an orientation (or index) for equilibria in bimatrix games. It is shown that the notion of orientation introduced here is the same as the notion of index introduced by Shapley (1974) (modulo the sign in the definition as explained in Remark 1.5). It is also shown that the L-H algorithm by Lemke and Howson (1964) that finds an equilibrium in a non-degenerate bimatrix game can be interpreted as a path-following algorithm in the dual construction. This allows one to visualise, in dimension 3 or lower, both the index and the L-H paths for all $m \times n$ non-degenerate bimatrix games with $\min\{m, n\} \leq 4$, whereas the interpretation of L-H paths and the definition of index by Shapley, or the interpretation by Savani and von Stengel (2004) by symmetrising games (see Section 1.3), uses geometric objects in dimension $m + n - 2$. Furthermore, it illustrates how non-degenerate bimatrix games fit into the study of solutions of piecewise linear equations as in Eaves and Scarf (1976).

This chapter is basic for the results in the subsequent chapters. Later, Chapter 3 shows how the results of this chapter are related to Sperner's Lemma in dimension $(m-1)$. In Chapter 4, the construction is used to give a strategic characterisation of the index in non-degenerate bimatrix games. Chapter 5 shows how the dual construction can be extended to outside option equilibrium components, which is applied in Chapter 6 to show that an outside option equilibrium component is hyperessential if and only if it has non-zero index.

The structure of this chapter is as follows. In Section 2.1 the dual construction is introduced and described in detail. Section 2.2 gives a characterisation of the Nash equilibria in the dual construction. Using only labels of player I, it is shown that the Nash equilibria are given by the fully labelled points in the dual construction (Proposition 2.6). Section 2.3 re-interprets the Lemke-Howson (L-H) algorithm and shows that it yields a connected path in the dual construction (Proposition 2.7 and Lemma 2.8). Finally, in Section 2.4, a notion of orientation for Nash equilibria is given. It is shown that it is equivalent to the notion of index defined by Shapley (Proposition 2.10).

2.1 The Dual Construction

This section describes a new geometric-combinatorial construction for non-degenerate bimatrix games. Put briefly, the subdivided strategy simplex X is dualised to obtain a dual space $|X^\triangle|$. Vertices in X become simplices in $|X^\triangle|$, and best reply regions in X become vertices in $|X^\triangle|$. There are two equivalent ways of constructing $|X^\triangle|$. One uses polar polytopes, the other one is a combinatorial dualisation method. Into $|X^\triangle|$ one then inscribes those faces of Y that are of strategic relevance for the game, yielding a division $X_*^\triangle$ of the dual space into labelled best reply regions for player I. The final construction has the same dimension as X and uses only labels of player I. The division into simplices reflects the best reply structure for player II, the division of the simplices into labelled best reply regions reflects the best reply structure for player I. Combining these two, the Nash equilibria are represented by completely labelled points in the dual construction.

The dual construction $|X^\triangle|$ can be obtained by using a polarisation method for polytopes (see e.g. Ziegler (1995, Section 2.3)). A combinatorial dualisation method is described further below. In brief, when polarising a polytope, vertices become simplices and facets become vertices. The polytope itself is obtained from the best reply polyhedron H in (1.5) that is given by the upper envelope of player II's expected payoffs over X. The polyhedron H is neither bounded nor full-dimensional. Since full-dimensional polytopes, i.e. bounded and full-dimensional polyhedra, are more convenient to study, the polyhedron H can be projected in order to obtain a polytope P that contains the same information as H and that is full-dimensional and bounded. This description is similar to von Stengel (2002), which also gives references to related earlier works.

The polyhedron H as in (1.5) is defined as

$$H = \{(x, v) \in \mathbb{R}^m \times \mathbb{R} \mid 1_m^\top x = 1, \ B^\top x \le 1_n v, \ x_i \ge 0 \ \forall \, i \in I\}.$$

Without loss of generality it can be assumed that $v > 0$ for all $(x, v) \in H$, since adding a positive constant to the entries of B does not affect the equilibria or the best reply structure of a game. Now consider the set

$$P' = \{x \in \mathbb{R}^m \mid B^\top x \le 1_n \ x_i \ge 0 \ \forall \, i \in I\}. \tag{2.1}$$

The mapping $H \to P' - \{0\}$ is given by $(x, v) \mapsto \frac{1}{v} \cdot x$, and the inverse $P' - \{0\} \to H$ is given by $x \mapsto \left(\frac{x}{|x|}, |x|\right)$, where $|x| = 1_m^\top x$. The vertex $\mathbf{0}$ of P' corresponds with "infinity" over H. The set P' is described by a finite number of inequalities and is both bounded and full-dimensional. Hence, the set P' is an m-dimensional polytope. Geometrically, the polytope P' is the projection of the polyhedron H on the hyperplane described by $v = 1$. This is depicted in Figure 2.1.

In order to obtain the polar (or dual) of a polytope of dimension m, it is convenient if $\mathbf{0} \in \mathbb{R}^m$ lies in the interior of the polytope. This is not the case for the polytope P', but can easily be obtained by translating the polytope P' to obtain the desired polytope P. Consider the point $\left(\frac{1}{m}, \ldots, \frac{1}{m}, \hat{v}\right) \in H$ with $\hat{v} = max_{i,j} b_{ij} + c$, where c is some arbitrarily large positive constant. The projection of this point is given by $\hat{x} = \left(\frac{1}{m\hat{v}}, \ldots, \frac{1}{m\hat{v}}\right) \in P'$ and lies in the interior of P'. So one can translate P' by $-\hat{x}$ to obtain

$$P = \{x \in \mathbb{R}^m \mid B^\top (x + \hat{x}) \le 1_n; \ x_i + \hat{x}_i \ge 0 \ \forall \, i \in I\}.$$

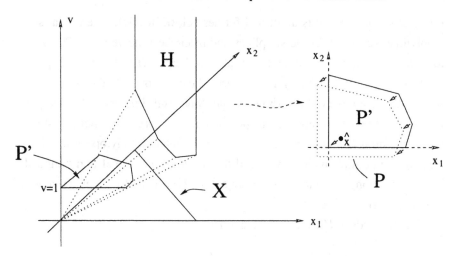

Fig. 2.1. The projection of the polyhedron H and the polytope P

Note that every other point in the interior of P' could be used for the translation. Then $\mathbf{0} \in \mathbb{R}^m$ lies in the interior of P. The polytope P is referred to as the *best reply polytope*. A depiction of P is given by the dotted lines on the right in Figure 2.1. The inequalities that describe P can be rewritten to obtain

$$P = \left\{ x \in \mathbb{R}^m \mid \frac{\hat{v}}{\hat{v} - \overline{B}_j} B_j^\top x \leq 1 \ \forall \ j \in N; \ -m\hat{v}x_i \leq 1 \ \forall \ i \in I \right\}, \qquad (2.2)$$

where $\overline{B}_j = \frac{\mathbf{1}_m^\top B_j}{m}$ is the average payoff for player II in column j.

In general, let P be a polytope given by

$$P = \left\{ z \in \mathbb{R}^m \mid c_k^\top z \leq 1, \ 1 \leq k \leq n \right\}.$$

Geometrically, the polytope P is defined by halfspaces, which are given by hyperplanes. The vectors $c_j \in \mathbb{R}^m$ are the normal vectors of these hyperplanes. The *polar* polytope $P^\triangle$ of the polytope P is defined as the convex hull of the normal vectors c_k of the hyperplanes that describe P, i.e.

$$P^\triangle = \text{conv} \left\{ c_1, \ldots, c_n \right\}. \qquad (2.3)$$

One can show that the polar of the polar polytope is the original polytope, i.e. $P^{\triangle\triangle} = P$ (see e.g. Ziegler (1995, Theorem 2.11)). Note that $\mathbf{0} \in \mathbb{R}^m$ lies in the interior of P, and hence in the interior of $P^\triangle$. A depiction of the polar polytope for a given polytope is given in Figure 2.2.

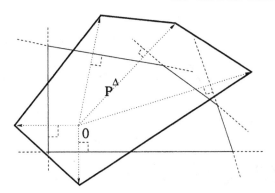

Fig. 2.2. The dual of a polytope

For a non-degenerate bimatrix game, the polytope P as in (2.2) is simple, i.e. each vertex of the m-dimensional polytope P is described by exactly m binding linear inequalities, so each vertex is contained in exactly m facets of P. Consequently, the polar $P^\triangle$ is simplicial (see e.g. Ziegler; Proposition 2.16). Each vertex of $P^\triangle$ corresponds to a facet of P, and each facet of $P^\triangle$, representing a vertex in P, is an $(m-1)$-simplex.

The study of polytopes is a very useful tool in the analysis of games. Von Stengel (1999b), for example, uses cyclic polytopes to construct games in order to obtain a new lower bound on the maximal number of Nash equilibria in a $d \times d$ non-degenerate bimatrix game. Savani and von Stengel (2004) employ a related method to construct games in which L-H paths are exponentially long.

The simplicial surface of the polar polytope $P^\triangle$ can be projected on the facet of $P^\triangle$ that is given by the $(m-1)$-simplex spanned by the vertices $-m\hat{v}e_i$, $i \in I$, where e_i denotes the unit vector in $\mathbb{R}^m$ with entry 1 in row i. The projection is defined by the intersection of the line between a point x and $(-m\hat{v})\mathbf{1}_m$ with the facet spanned by $-m\hat{v}e_i$, $i \in I$ (see Figure 2.3). This yields a triangulation of the facet spanned by the vertices $-m\hat{v}e_i$, $i \in I$. A *triangulation* (or *simplicial subdivision*) of a simplex is a finite collection of smaller simplices whose union is the simplex, and that is such that any two of the simplices intersect in a face common to both, or the intersection is empty. The vertices of a triangulation are the vertices of the simplices in the triangulation.

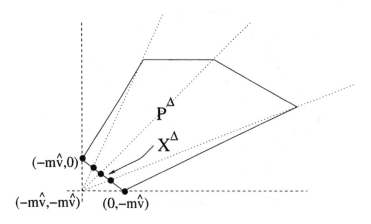

Fig. 2.3. The simplicial division of $X^\triangle$

Definition 2.1. *The simplex spanned by $-m\hat{v}e_i$, $i \in I$, is denoted as $X^\triangle$. The triangulation induced by the projection $P^\triangle - X^\triangle \to X^\triangle$ is denoted as $|X^\triangle|$, and referred to as the dual construction. The facets of $P^\triangle$ other than $X^\triangle$, which are $(m-1)$-simplices, are denoted as $v^\triangle$. For notational parsimony, their projections on $X^\triangle$, which are also $(m-1)$-simplices, are also denoted as $v^\triangle$.*

An illustration of $|X^\triangle|$ is depicted in Figure 2.3. The vertices $-m\hat{v}e_i$ correspond to the facets of P that represent unplayed strategies. All other vertices of $P^\triangle$ correspond to facets of P that represent best reply facets of H. Each vertex $v \neq -\hat{x}$ of P represents a vertex of H, and hence a vertex in the division of X into best reply regions. So each vertex v in X or H corresponds to a unique $(m-1)$-simplex $v^\triangle$ in $|X^\triangle|$ or on the surface of $P^\triangle$. The simplex $X^\triangle$ represents the vertex $-\hat{x} \in P$, and is spanned by $-m\hat{v}e_i$, $i \in I$.

   The induced triangulation $|X^\triangle|$ is regular. A triangulation is called *regular* if it arises as the projection of a polytope Q whose facets are simplices (see e.g. Ziegler (1995, Definition 5.3)). The simplices in $|X^\triangle|$ are the projections of the facets of $P^\triangle$. Essentially, the projection $|X^\triangle|$ is a so-called *Schlegel-diagram* of $P^\triangle$ that is combinatorially equivalent to the complex $\partial P^\triangle - X^\triangle$ (see e.g. Ziegler (1995, Proposition 5.6.)), where $\partial P^\triangle$ denotes the boundary of $P^\triangle$.

Now suppose one has a regular triangulation $|X^\triangle|$ of $X^\triangle$. Assume that the only vertices of the triangulation that lie on the boundary of $X^\triangle$ are those that span $X^\triangle$, i.e. $-m\hat{v}e_i$, $i \in I$. Then one can obtain a payoff matrix B that induces this subdivision. For this, consider the polytope Q that induces this triangulation. Without loss of generality it can be assumed that $\mathbf{0} \in Q$. Otherwise the vectors other than $-m\hat{v}e_i$, $i \in I$, can be moved in the same manner along the projection line. Then Q is the polar polytope $P^\triangle$ of a polytope P. The polytope $P^\triangle$ is given by $\mathrm{conv}\{c_1,\ldots,c_n\}$ (see (2.3)), where the first m vectors are given by $-m\hat{v}e_i$, $i \in I$ (these are the vertices of $X^\triangle$). Given a polytope $P^\triangle$, the following lemma shows how one can construct the corresponding payoff matrix B that yields $P^\triangle$ as the polar of the polytope P given in (2.2).

Lemma 2.2. *Consider $P^\triangle$ as in (2.3) with $\mathbf{0} \in P^\triangle$, and let the first m vectors be given by $c_i = -m\hat{v}e_i$, $i \in I$. For all other c_j, $j > m$, let $(c_j)_i > -m\hat{v}\ \forall\, i \in I$, where $(c_j)_i$ denotes the i-th row of c_j, and let $\bar{c}_j > -\hat{v}$, where $\bar{c}_j = \frac{\mathbf{1}_m^\mathsf{T} c_j}{m}$. Then $P^\triangle$ is the polar of the polytope in (2.2) with*

$$B_j = \frac{\hat{v}}{\hat{v}+\bar{c}_j}c_j. \tag{2.4}$$

Proof. By definition, one has $\frac{\hat{v}}{\hat{v}-\bar{B}_j}B_j = c_j$ for all $j > m$. This implies that $\frac{\hat{v}}{\hat{v}-\bar{B}_j}\bar{B}_j = \bar{c}_j$, so $\bar{B}_j = \frac{\hat{v}}{\hat{v}+\bar{c}_j}\bar{c}_j$. Substituting this into $B_j = \left(\frac{\hat{v}-\bar{B}_j}{\hat{v}}\right)c_j$ yields $B_j = \frac{\hat{v}}{\hat{v}+\bar{c}_j}c_j$. Note that the first m vectors are $c_i = -m\hat{v}e_i$, $i \in I$, and give the inequalities $-m\hat{v}x_i \le 1$ in (2.2).

Translating P as in (2.2) by $(\frac{1}{m\hat{v}},\ldots,\frac{1}{m\hat{v}})$ gives the polytope P' as in (2.1) with $(\frac{1}{m\hat{v}},\ldots,\frac{1}{m\hat{v}})$ lying in the interior of P'. From $P' - \{\mathbf{0}\}$ one obtains H via $x \mapsto \left(\frac{x}{|x|},|x|\right)$. So the upper envelope H satisfies $v > 0$ for all $(x,v) \in H$, and $(\frac{1}{m},\ldots,\frac{1}{m},\hat{v})$ lies in the relative interior of H with $\hat{v} > \bar{B}_j\ \forall\, j \in N$. $\square$

The above construction shows that each strategy simplex X can be dualised in a way such that one obtains a regular triangulation $|X^\triangle|$ of an $(m-1)$-simplex. This construction is such that the vertices of X correspond to the simplices in $|X^\triangle|$, and the best reply regions and unplayed strategies in X correspond to vertices in $|X^\triangle|$. Furthermore, an edge in X that connects vertices v_1 and v_2 in X corresponds to the common $(m-2)$-face of the two adjacent $(m-1)$-simplices $v_1^\triangle$ and $v_2^\triangle$ in $|X^\triangle|$.

The important aspects of $|X^\triangle|$ are the combinatorial properties of the simplices and vertices in $|X^\triangle|$. A combinatorial equivalent of $|X^\triangle|$, which, for notational parsimony, is also referred to as $|X^\triangle|$, can be obtained without using the polarisation method from above. Instead, it can be derived directly from the division of X into best reply regions. To illustrate the procedure, it is applied to the following example.

Example 2.3.

$$\begin{bmatrix} 0,0 & 10,10 & 0,0 & 10,-10 \\ 10,0 & 0,0 & 0,10 & 0,8 \\ 8,10 & 0,0 & 10,0 & 8,8 \end{bmatrix} \tag{2.5}$$

Take player I's standard $(m-1)$-simplex representing the mixed strategy space X. Then X can be divided into best reply regions $X(j)$. Non-degeneracy implies that the number of best replies in a vertex $v \in X$ equals the number of strategies played with positive probability in v. Figure 2.4 gives the division of X into best reply regions for player II for the game in Example 2.3. It shows that every vertex $v \in X$ has exactly m labels, where the labels of a vertex $v \in X$ are the pure best reply strategies of player II with respect to v and the pure strategies of player I not played in v. The labels of a point $x \in X$ are given by $L(x)$ as defined in (1.3).

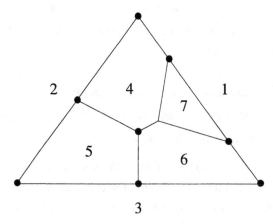

Fig. 2.4. The best-reply division of X for the game in Example 2.3

A combinatorial dualisation of X is now obtained as follows. For each best reply region and each unplayed strategy, one chooses a representative point in

$\mathbb{R}^{m-1}$ that serves as a vertex in $|X^\triangle|$. For best reply regions, these representatives are denoted as $X(j)^\triangle$. For an unplayed strategy $i \in I$ the representatives are denoted as $X(i)^\triangle$.

The points $X(k)^\triangle$, for $k \in I \cup J$, that are corresponding to best reply regions or unplayed strategies, now become the vertices in the dual of X, so each such vertex has label k. For every vertex $v \in X$ with labels $L(v)$, the combinatorial dual simplex $v^\triangle$ is the simplex spanned by the dual vertices $X(k)^\triangle$, with $k \in L(v)$. For two vertices v_1 and v_2 that are joined by an edge with labels $L(v_1) \cap L(v_2)$ in X, the two combinatorial simplices $v_1^\triangle$ and $v_2^\triangle$ are adjacent and share the $(m-2)$-face that is spanned by the dual vertices representing the labels $L(v_1) \cap L(v_2)$ in $X^\triangle$.

For the game in Example 2.3, the triangulation $|X^\triangle|$ is illustrated in Figure 2.5. The dotted lines in Figure 2.5 show the division of X into best reply regions. The solid lines illustrate $|X^\triangle|$. The best reply regions in X and those labels that represent unplayed strategies become dual vertices in $|X^\triangle|$. Each vertex in X is represented by a unique $(m-1)$-simplex in $|X^\triangle|$. The edges in X become $(m-2)$-faces of two adjacent simplices in $|X^\triangle|$.

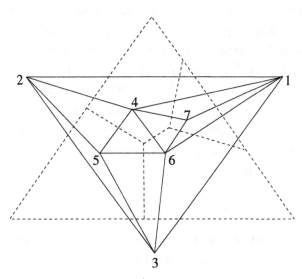

Fig. 2.5. The triangulation of $X^\triangle$ for Example 2.3

If a vertex of a simplex $v^\triangle$ is of the form $X(i)^\triangle$, for some $i \in I$, it is called an *outer vertex* of $v^\triangle$. Outer vertices of $v^\triangle$ represent those strategies of player I that are played with zero probability in v. The $(m-1)$-simplex $X^\triangle$ is spanned by all outer vertices $X(i)^\triangle$, $i \in I$. Accordingly, the *inner vertices* of a simplex $v^\triangle$ are of the form $X(j)^\triangle$, for some $j \in J$. The inner vertices of a simplex $v^\triangle$ represent best reply strategies of player II. All simplices $v^\triangle$ have at least one inner vertex, simplices representing a pure strategy of player I have exactly one inner vertex.

2.2 Labelling and Characterisation of Nash Equilibria

The aim is now to divide the simplex $X^\triangle$ into regions with labels $i \in I$ such that the Nash equilibria are represented by fully labelled points. As above, it can be assumed that all entries of the payoff matrix A are strictly greater than zero. Now consider a simplex $v^\triangle \in |X^\triangle|$. An inner vertex that represents the pure strategy of $j \in N$ of player II has the corresponding payoff column A_j. The outer vertices do not represent payoff columns of A and are dealt with by introducing slack variables. Each outer vertex that represents a pure strategy $i \in I$ of player I played with zero probability is assigned an *artificial payoff vector* e_i, i.e. the unit vector in $\mathbb{R}^m$ with entry 1 in row i. So suppose $I(v) = \{i_1, \ldots, i_k\}$, so $v^\triangle$ is spanned by outer vertices $X(i_1)^\triangle, \ldots, X(i_k)^\triangle$ and some inner vertices $X(j_{k+1})^\triangle, \ldots, X(j_m)^\triangle$. The payoffs for player I with respect to pure strategies $j_{k+1}, \ldots, j_m$ are given by the columns $A_{j_{k+1}}, \ldots, A_{j_m}$ of the payoff matrix A. The artificial payoffs for player I with respect to the unplayed strategies $i_1, \ldots, i_k$ are defined as $e_{i_1}, \ldots, e_{i_k}$. Let $A(v)$ be the following *artificial payoff matrix*,

$$A(v) = \begin{bmatrix} e_{i_1} & \cdots & e_{i_k} & A_{j_{k+1}} & \cdots A_{j_m} \end{bmatrix}. \tag{2.6}$$

This artificial payoff matrix now allows one to divide each simplex $v^\triangle$ into labelled "best reply" regions with labels $i \in I$.

Definition 2.4. *A point in $v^\triangle$ is denoted as w_s, described by its convex coordinates with respect to the vertices of $v^\triangle$ (the subscript "s" indicates that w_s contains slack variables).*

Then every simplex $v^\triangle$ can be divided into labelled regions according to

$$v^\triangle(i) = \{w_s \in v^\triangle \mid (A(v)w_s)_i \geq (A(v)w_s)_k \ \forall \ k \in I\}. \qquad (2.7)$$

This is the same division as the division of player II's mixed strategy space in the case $A(v)$ is the payoff matrix of player I in some bimatrix game.

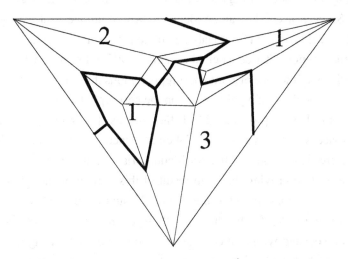

Fig. 2.6. The labelled dual construction $X_*^\triangle$ for Example 2.3

Dividing each simplex $v^\triangle$ in $|X^\triangle|$, this gives, by non-degeneracy, a division of $X^\triangle$ into full-dimensional regions $X^\triangle(i)$ with labels $1,\ldots,m$, where

$$X^\triangle(i) = \bigcup_{v \in V} v^\triangle(i).$$

This division is well-defined, since, if two simplices $v_1^\triangle$ and $v_2^\triangle$ share some common face, the induced division on that face is the same in both simplices $v_1^\triangle$ and $v_2^\triangle$. For the game in Example 2.3 the resulting division of $X^\triangle$ is depicted in Figure 2.6.

Definition 2.5. *The division of $X^\triangle$ into labelled regions $X^\triangle(i)$ is referred to as the labelled dual construction, and is denoted as $X_*^\triangle$. A point $w_s \in X_*^\triangle$ is assigned the labels $I(w_s)$ of those regions that contain w_s, i.e.*

$$I(w_s) = \{i \in I \mid w_s \in X^\triangle(i)\}. \qquad (2.8)$$

For each simplex $v^\triangle$, the inner $k+1$ (for some $k \geq 0$) vertices of $v^\triangle$ span some k-face of $v^\triangle$. This k-face is referred to as the *best reply face* of $v^\triangle$ and is denoted as $v^{\mathrm{br}\triangle}$. So the best reply face $v^{\mathrm{br}\triangle}$ is spanned by exactly those vertices of $v^\triangle$ that represent a best reply strategy of player II with respect to strategy v. The best reply face $v^{\mathrm{br}\triangle}$ corresponds to the face of Y that is spanned by those pure strategies of player II that are represented as vertices of $v^{\mathrm{br}\triangle}$. So each $w \in v^{\mathrm{br}\triangle}$ can be identified with a unique strategy $y \in Y$ of player II. The division of $v^\triangle$ into labelled regions also yields a division of $v^{\mathrm{br}\triangle}$ into labelled regions. These labelled regions are affine linear transformations of the division of the face of Y into best reply regions that corresponds to $v^{\mathrm{br}\triangle}$. It should be noted that if a point w lies on the best reply face of a simplex $v^\triangle$, then the set of labels $I(w)$ as in (2.8) is the same as $I(w)$ in (1.1).

The space $X_*^\triangle$ together with the labelling function in (2.8) now allows a complete characterisation of the Nash equilibria of a non-degenerate bimatrix game. Before proving the main result of this section, it should be noted that all points w_s that lie in the interior of $X^\triangle$ and in some $v^\triangle$ can be projected on some $w \in v^{\mathrm{br}\triangle}$ by dropping those coordinates that are the slack variables associated with artificial payoff vectors and normalising the resulting vector such that its entries sum to 1. So let $w_s \in v^\triangle$. Let the set of outer vertices of $v^\triangle$ be $X(i_1)^\triangle, \ldots, X(i_k)^\triangle$, and let the set of inner vertices of $v^\triangle$ be $X(j_{k+1})^\triangle, \ldots, X(j_m)^\triangle$. Note that for all simplices $v^\triangle$, the set of inner vertices is non-empty. So let $w_s = (w_{s1}, \ldots, w_{sm})$, where the first k entries are the coordinates with respect to the outer vertices, and the last $m-k$ entries are the coordinates with respect to the inner vertices. Then define the projection $p(w_s)$ as

$$w = p(w_s) = \begin{cases} w_i = 0 & ; \quad 1 \leq i \leq k \\ w_i = \frac{w_{si}}{\sum_{i=k+1}^m w_{si}} & ; k+1 \leq i \leq m \end{cases} \tag{2.9}$$

The projection point $w = p(w_s) \in v^{\mathrm{br}\triangle}$ can be identified with a unique strategy vector in Y. For w_s on the boundary of $X_*^\triangle$, one defines $p(w_s) = \mathbf{0} \in \mathbb{R}^m$. This allows the following characterisation.

Proposition 2.6. *A point* $w_s \in X_*^\triangle$ *with* $w_s \in v^\triangle$ *is completely labelled if and only if* $(v, p(w_s))$ *is a Nash equilibrium of the game.*

Proof. Let w_s be completely labelled with $w_s \in v^\triangle$. Then consider the artificial payoff matrix $A(v)$. A point is, by definition, completely labelled if

$A(v)w_s = c\mathbf{1}_m$, where c is some positive constant. It is easy to verify that the payoffs of $A(v)$ are non-degenerate, since the payoffs of A are non-degenerate. Hence w_s lies in the interior of $v^\triangle$. By construction one has $w = p(w_s) \in v^{\mathrm{br}\triangle}$. It implies that $I(w) = I - I(v)$, where $I(v)$ is as defined in (1.2). Since w lies on the best reply face of $v^\triangle$, it means that player II mixes only those strategies with positive probability in w that are a best reply to v. So, using (1.1) and (1.2), one has

$$w \in v^{\mathrm{br}\triangle} \iff J(v) \cup J(w) = J. \tag{2.10}$$

This is to say that player II is always in equilibrium when considering points in the labelled dual construction. But then $I(w) = I - I(v)$, so $I(v) \cup I(w) = I$. This means that (v, w) is completely labelled, and hence an equilibrium.

Now let (v, w) be a Nash equilibrium. Then $J(v) \cup J(w) = J$, so $w \in v^{\mathrm{br}\triangle}$. Since it is a Nash equilibrium, one has $I(v) = I - I(w)$. So $A(v)w$ is a vector with maximum entries in those rows that are strategies played with positive probability in v. Let c be this maximum entry. Now assign weights to the columns representing unplayed strategies to obtain a strictly positive vectors $\tilde{w}_s$ such that $A(v)\tilde{w}_s = c\mathbf{1}_m$. Normalising the vector $\tilde{w}_s$ such that the entries add up to one yields the desired vector w_s with $I(w_s) = I$.   □

For the game in Example 2.3, the labelled dual construction is depicted in Figure 2.6. For the following description, the coordinates of w_s carry a subscript, marking the payoff vector they apply to. So, for example, the subscripts $1, 2, 3$ refer to artificial payoff vectors, and the subscripts $4, 5, 6, 7$ refer to payoff columns of A. The construction contains three completely labelled points, namely $w_s = ((\frac{8}{9})_1, (\frac{8}{90})_4, (\frac{2}{90})_7)$ lying in the simplex $v^\triangle$ representing $v = (0, \frac{1}{5}, \frac{4}{5})$, the point $w_s' = ((\frac{5}{11})_4, (\frac{5}{11})_5, (\frac{1}{11})_6)$ lying in the simplex representing $v' = (\frac{1}{3}, \frac{1}{3}, \frac{1}{3})$, and $w_s'' = ((\frac{10}{21})_2, (\frac{10}{21})_3, (\frac{1}{21})_5)$ lying in the simplex representing $v'' = (1, 0, 0)$. Projecting these vectors gives $w = (\frac{4}{5}, 0, 0, \frac{1}{5})$, the point $w' = (\frac{5}{11}, \frac{5}{11}, \frac{1}{11}, 0)$ and $w'' = (0, 1, 0, 0)$. So (v, w), (v', w') and (v'', w'') are the Nash equilibria of the game.

Instead of labelling the dual construction $|X^\triangle|$, which consists of the projected simplicial facets of the polar polytope $P^\triangle$, one can also label the simplicial facets of $P^\triangle$ directly via the artificial payoff matrix. The division of each simplicial facet of $P^\triangle$ is obtained in the same way as the division of the projected simplices. The result of this construction is depicted in Figure 2.7

for the game given by the payoff matrices

$$A = \begin{bmatrix} 1 & 0 & 0 \\ 0 & 1 & 1 \end{bmatrix}; B = \begin{bmatrix} 6 & 4 & 1 \\ 1 & 3 & 5 \end{bmatrix}.$$

The resulting labelled surface of the polar polytope is denoted as $P_*^{\triangle}$. Its simplicial surface is denoted as $|P^{\triangle}|$. In this construction, the equilibria are, as before, represented by exactly those points on the surface of the polar polytope that are completely labelled. The artificial equilibrium $(\mathbf{0}, \mathbf{0})$ can be identified with the completely labelled point on the facet $X^{\triangle}$ of $P_*^{\triangle}$. Note that $X^{\triangle}$ corresponds to the vertex of P' that has all labels of player I, i.e. no strategy of player I is played with positive probability. So the artificial payoff matrix that corresponds to this facet is the identity matrix that only consists of artificial payoff vectors. Its centre is a completely labelled point. So, instead of considering the projection of the labelled facets, one might as well characterise the equilibria using the "labelled sphere" $P_*^{\triangle}$.

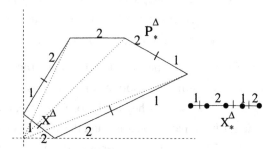

Fig. 2.7. The labelled polar polytope $P_*^{\triangle}$

The labelled dual construction allows one to completely characterise the Nash equilibria of a non-degenerate bimatrix game in a geometric object of dimension $m - 1$ by using only the set I of labels of player I. Assuming without loss of generality $m \leq n$, it is possible to visualise $X_*^{\triangle}$ for all $m \leq 4$. It also demonstrates how non-degenerate bimatrix games fit into the study of solutions of piecewise linear equations as in Eaves and Scarf (1976), and allows one to illustrate how one can find a Nash equilibrium of a non-degenerate bimatrix game.

2.3 The Lemke-Howson Algorithm in the Labelled Dual Construction

The L-H algorithm described in Section 1.2 is the standard algorithm for finding a Nash equilibrium in a non-degenerate bimatrix game. The L-H algorithm describes a path in the product space $X \times Y$ (or $X_0 \times Y_0$ when including the artificial equilibrium points) that is given by a set of points $(x,y) \in X \times Y$ that is described by labels $L(x) \cup L(y) = I \cup J - \{k\}$ for some $k \in I \cup J$. This path consists of pairs of edges and vertices in the product graph.

The fact that the L-H algorithm applies to a product graph makes it difficult to visualise it for games of higher dimension. In this section, it is shown that every L-H path in $X \times Y$ that is defined by a missing label $k \in I$ of player I can be interpreted as a path in the labelled dual $X_*^{\triangle}$ that consists of paths that are almost completely labelled with missing label k. This allows one to give a new geometric interpretation not only of the L-H algorithm but also of the fact that equilibria at the ends of an L-H path have opposite indices (see Section 2.4 below).

Similar to the definition of $M(k)$ in (1.9), one can define the set of almost completely labelled points on the labelled surface $P_*^{\triangle}$ for a missing label k of player I. So let $M(k)_*^{\triangle}$, for $k \in I$, denote all those points w_s in $P_*^{\triangle}$ that have at least labels $I - \{k\}$, i.e.

$$M(k)_*^{\triangle} = \{w_s \in P_*^{\triangle} \mid I - \{k\} \subset I(w_s)\}. \tag{2.11}$$

One obtains the following proposition (compare Theorem 1.3).

Proposition 2.7. *Let G be a non-degenerate $m \times n$ bimatrix game. Fix a label $k \in I$. Then $M(k)_*^{\triangle}$ consists of disjoint paths and cycles in $P_*^{\triangle}$. The endpoints are the equilibria of the game, including the artificial equilibrium.*

Proof. As before, let $|P^{\triangle}|$ denote the simplicial surface of $P^{\triangle}$. Since the payoff matrix $A(v)$ is non-degenerate for all simplices $v^{\triangle}$ in $|P^{\triangle}|$, the set of almost completely labelled points in $v^{\triangle}$ with a missing label k is, if not empty, an edge (or line segment) in $v^{\triangle}$. Now take an endpoint $w_s \in v^{\triangle}$ of an edge in $v^{\triangle}$ with labels $I - \{k\}$. Then there are two cases. The first is where w_s lies in the interior of $v^{\triangle}$. In this case, w_s represents an equilibrium and is fully labelled. So w_s is endpoint of a unique edge in $v^{\triangle}$. The second case is where w_s lies on the boundary of $v^\triangle$. In this case, due to the non-degeneracy assumption, the point w_s lies in the interior of some $(m-2)$-face of $v^\triangle$. This $(m-2)$-face is the face of another simplex $v'^\triangle$ in $|P^\triangle|$ that is adjacent to $v^\triangle$. In $v'^\triangle$, the point w_s must be the endpoint of another edge with labels $I - \{k\}$. So the endpoints of edges of $M(k)^\triangle_*$ in $v^\triangle$ are incident to one or two edges of $M(k)^\triangle_*$ in $P^\triangle_*$. □

Note that $X^\triangle_*$ is just a projection of the labelled facets of $P^\triangle_* - X^\triangle$ on $X^\triangle$. So the paths and cycles in $X^\triangle_*$ with labels $I - \{k\}$ are projections of the paths and cycles in $P^\triangle_* - X^\triangle$ with labels $I - \{k\}$. For notational convenience, the projection of these paths and cycles in $X^\triangle_*$ is also denoted as $M(k)^\triangle_*$. Equivalently, one can define $M(k)^\triangle_* = \{w_s \in X^\triangle_* \mid I - \{k\} \subset I(w_s)\}$. The endpoints of the paths in $X^\triangle_*$ are the equilibria of the game, not including the artificial equilibrium, since the artificial equilibrium lies on the face $X^\triangle$ on which $P^\triangle_* - X^\triangle$ is projected. I.e. the artificial equilibrium is not seen under the projection and can be thought of lying under $X^\triangle_*$. In the same way as above one can confirm that $M(k)^\triangle_*$ in $X^\triangle_*$ consists of paths and cycles.

The following lemma shows how the definitions of $M(k)$ and $M(k)^\triangle_*$ are related. This yields a straightforward interpretation of the L-H algorithm on the labelled surface $P^\triangle_*$ and in the labelled dual construction $X^\triangle_*$.

Lemma 2.8. *Equilibria that are connected by a L-H path in $M(k)$ are connected by a path in $M(k)^\triangle_*$. An edge $e_X \times \{w\} \in M(k)$ is represented in $M(k)^\triangle_*$ by two adjacent simplices. An edge $\{v\} \times e_Y \in M(k)$ is represented in $M(k)^\triangle_*$ by an edge in $v^\triangle$ with labels $I - \{k\}$.*

Proof. First consider an edge $e_X \times \{w\} \in M(k)$. Then e_X is an edge in X_0. Let this be an edge in X between v_1 and v_2. Edges in X_0 are represented in $|X^\triangle|$ and $|P^\triangle|$ by an $(m-2)$-face that is common to $v^\triangle_1$ and $v^\triangle_2$. As for the edge that connects the artificial equilibrium with a pure strategy, i.e. the edge between $\mathbf{0}$ and a pure strategy v, note that every pure strategy v is represented in $|P^\triangle|$ by a simplex $v^\triangle$ that is adjacent to $X^\triangle$, the latter representing the artificial strategy $\mathbf{0} \in \mathbb{R}^m$. In $X^\triangle_*$ this is reflected by the fact that $v^\triangle$ has an $(m-2)$-face on the boundary of $X^\triangle_*$. So, if (v_1, w) and (v_2, w) lie along a L-H path, then $v^\triangle_1$ and $v^\triangle_2$ are adjacent and share the $(m-2)$-face that corresponds to the labels $L(v_1) \cap L(v_2)$. So the L-H path in X_0 yields a union of adjacent simplices in $|X^\triangle|$ and $|P^\triangle|$.

Now suppose one has $(v, w) \in M(k)$. Let $(v, w) \in X \times Y$. Then, by the equivalence in (2.10), one has $w \in v^{\mathrm{br}\triangle}$. This point corresponds to an almost completely labelled point $w_s = l(w) \in v^{\triangle}$ in the labelled dual construction. To see this, let $(w_s)_k$, $k \in I(v) \cup J(v)$, denote the row of w_s that corresponds to the column of $A(v)$ that represents strategy k. Also, let w_k, $k \in J(v)$, denote the probability with which strategy k is played in w. Then define

$$\tilde{l}(w)_k = \begin{cases} w_k & k \in J(v) \\ c - (Aw)_k & k \in I(v) \end{cases},$$

where c is the maximum payoff for player I when player II plays w, and $(Aw)_k$ is the payoff for player I in strategy k. In v, a strategy $k \in I(v)$ has probability zero. So, for $k \in I(v)$, the expected payoff for the unplayed strategy k is $(Aw)_k$. Normalising $\tilde{l}(w)$ yields the vector $w_s = l(w)$ such that $I(w_s) = I(v) \cup I(w)$, so $w_s \in M(k)_*^{\triangle}$. Therefore, the mapping $l(w)$ is a lifting of $w \in v^{\mathrm{br}\triangle}$ to a point $w_s \in v^{\triangle}$ such that $I(w_s) = I(v) \cup I(w)$ (compare the projection p in (2.9)).

Now consider an edge $\{v\} \times e_Y \in M(k)$ that connects (v, w_1) and (v, w_2) with $w_1 \neq 0$ and $w_2 \neq 0$. By the equivalence in (2.10) one sees that then $e_Y \subset v^{\mathrm{br}\triangle}$, so the edge lies on the best reply face of $v^{\triangle}$. But that means that $l(e_Y)$ is an edge in $v^{\triangle}$ connecting $l(w_1)$ and $l(w_2)$.

It remains to show that these lifted edges yield a connected path in the union of simplices that correspond to the L-H path in X_0. So let w be an endpoint of the edge e_Y. Then one can distinguish two cases.

The first is where $I(v) \cap I(w) = \{i\}$. In this case the pair (v, w) has a duplicate label i of player I. This means that strategy i of player I is a best reply, but is not played with positive probability in v. Therefore, one has $(Aw)_i = c$, so $l(w)_i = 0$, i.e. the lifted point $l(w)$ lies on the $(m-2)$-face where the weight on the artificial payoff vector e_i is zero. So it lies on the $(m-2)$-face that corresponds to labels $L(v) - \{i\}$. This represents the edge in X_0 that is described by labels $L(v) - \{i\}$ and connects v and another vertex v', with (v, w) and (v', w) both lying along a L-H path in $M(k)$. So the lifted point is adjacent to two edges, one in $v^{\triangle}$ and one in $v'^{\triangle}$.

The second case is where $I(v) \cap I(w) = \emptyset$. In this case (v, w) has a duplicate label j of player II. This implies that strategy j of player II is a best reply, but is not played with positive probability. Therefore, $w_j = 0$ and hence $l(w)_j = 0$, i.e. the lifted point $l(w)$ lies on the $(m-2)$-face of $v^{\triangle}$ where the weight on the payoff vector A_j is zero. So it lies on the $(m-2)$-face that corresponds to labels $L(v) - \{j\}$. This represents the edge in X_0 that is described by labels $L(v) - \{j\}$ and connects v and another vertex v', with (v,w) and (v',w) both lying along a L-H path in $M(k)$. So the lifted point is also adjacent to two edges, one in $v^\triangle$ and one in $v'^\triangle$.

Finally, one has to account for the simplices adjacent to $X^\triangle$ and the artificial equilibrium. The L-H path with missing label k that starts in the artificial equilibrium is such that, after two steps, it yields the pair (v,w), where v represents pure strategy k, and w is the pure best reply to v. Then either (v,w) is an equilibrium, in which case the completely labelled point in $v^\triangle$ is connected with the completely labelled point in $X^\triangle$ via an edge in $v^\triangle$ and an edge in $X^\triangle$. If (v,w) is not an equilibrium, pure strategy v is not a best reply to pure strategy w. The lifted point $l(w)$ lies on the $(m-2)$-face of $v^\triangle$ that corresponds to labels $L(v) - I(w)$, and is also connected with the completely labelled point in $X^\triangle$ via an edge in $v^\triangle$ and an edge in $X^\triangle$. For pure strategies v and w such that (v,w) is an equilibrium, the completely labelled point w_s in $v^\triangle$ connects with a point on the $(m-2)$-face corresponding to labels $L(v) - \{k\}$. This is also the $(m-2)$-face of $v'^\triangle$ such that (v,w) and (v',w) both lie along a L-H path in $M(k)$. $\qquad\square$

The above lemma can be illustrated by considering the paths $M(2)^\triangle_*$ for the game in Example 2.3. This is depicted in Figure 2.8. According to the L-H algorithm, one starts at the artificial equilibrium $v_0 = \mathbf{0}, w_0 = \mathbf{0}$ and looks at the path that has labels $1,3$. Dropping label 2 means that one flips from the artificial equilibrium simplex $X^\triangle$ into the simplex $v_1^\triangle$ that represents pure strategy 2 of player I. Then v_1 has labels $1,3,6$, since 6 is a best reply to pure strategy 2, and w_0 has labels $4,5,6,7$. Hence 6 is a duplicate label. This determines w_1. Strategy w_1 represents the pure best reply to pure strategy 2, which is 6. So $w_1 = (0,0,1,0)$ with labels $4,5,7,3$, since pure strategy 3 is a best reply to w_1. In $X^\triangle_*$, this is represented by w_{s1}. Now 3 is a duplicate label. This determines the simplex $v_2^\triangle$ by flipping over the face that corresponds to vertices representing strategies 1 and 6. Then v_2 has labels $1,7,6$. Now 7 is a duplicate label, determining w_2. The strategy w_2 is the mixed strategy that mixes strategies 6 and 7, with best replies 1 and 3. In $X^\triangle_*$, this gives w_{s2}. Now w_2 has labels $5,4,1,3$, so 1 is a duplicate label, which determines $v_3^\triangle$. The

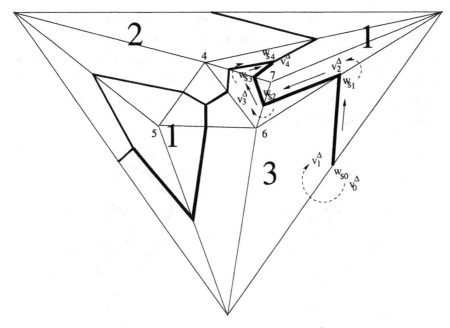

Fig. 2.8. The L-H paths for $k = 2$ in $X_*^\triangle$

simplex $v_3^\triangle$ is the simplex adjacent to $v_2^\triangle$ with common face spanned by vertices representing 6 and 7. This is the simplex spanned by vertices representing 4,6,7. Now 4 is duplicate, which determines w_3 in which pure strategy 4 is played with positive probability. In $X_*^\triangle$, this gives w_{s3}. Strategy w_3 has labels 4,6,1,3, so now 6 is a duplicate label. Flipping over the face of $v_3^\triangle$ that is spanned by vertices 4 and 7 gives $v_4^\triangle$ spanned by vertices representing by 4,7 and 1. Finally, label 1 is duplicate, determining w_4 with labels 5,6,2,3, which, in $X_*^\triangle$, is represented w_{s4}. The tuple (v_4, w_4) is an equilibrium of the game.

This reinterpretation of the L-H paths in $X_*^\triangle$ also allows one to illustrate why Nash equilibria might be inaccessible in the sense that they are not connected via a union of paths with the artificial equilibrium as noted by Shapley (1974). An example for this situation is depicted on the left in Figure 2.9. The union of paths $M_*^\triangle(k)$, for $k \in I$, is depicted in bold lines. The game represented on the left in Figure 2.9 has three equilibria, one pure strategy equilibrium and two in which player I plays all three strategies with positive probability. Starting at one mixed strategy equilibrium, every path in $M_*^\triangle(k)$

always leads to the other mixed strategy equilibrium and vice versa. So for $k \in I$, the L-H algorithm only finds the pure strategy equilibrium in which player I plays only pure strategy 1 (the equilibria might not be isolated when considering paths $M(j)$ for $j \in J$). $X_*^{\triangle}$ can also be used to show that $M_*^{\triangle}(k)$ might contain cycles. This is depicted on the right in Figure 2.9, which illustrates a cycle with labels $1, 3$ in $M_*^{\triangle}(2)$.

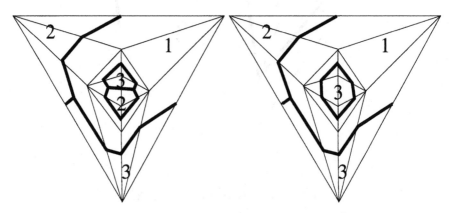

Fig. 2.9. Inaccessible equilibria and cycles in $X_*^{\triangle}$

2.4 An Orientation for Nash Equilibria

This section gives a re-interpretation of the index by means of the labelled dual construction. This allows a simple visualisation of the index for any $m \times n$ bimatrix game with $m \leq 4$, since $X_*^{\triangle}$ is of dimension $m - 1$ for an $m \times n$ bimatrix game. Furthermore, this re-interpretation of the index extends to certain components of equilibria, namely outside option equilibrium components in bimatrix games (Chapter 5). This re-interpretation of the index is then employed in Chapter 4 to obtain a strategic characterisation of the index in non-degenerate bimatrix games and in Chapter 6 to obtain a characterisation of hyperessentiality in terms of the index.

   The definition of the index in $X_*^{\triangle}$ is similar to the index as depicted in Figure 1.5, i.e. it is defined by the relative ordering of the labels "around" an equilibrium. Consider a completely labelled point $w_s \in X_*^{\triangle}$ that represents an equilibrium. Note that in this case w_s lies in the interior of some unique $v^\triangle$. One now constructs a simplex $w_s^\triangle$ such that it contains w_s and such that each vertex of $w_s^\triangle$ lies in a different best reply region of $v^\triangle$. Comparing the orientation of this simplex with the orientation induced by $X^\triangle$ then yields the index of the equilibrium represented by w_s.

The simplex $w_s^\triangle$ can be obtained as follows. Let $w_s \in v^\triangle$ be completely labelled. For $i \in I$, let w_i denote the vector, described as a convex combination of the vertices of $v^\triangle$, such that the payoff for player I from the artificial payoff matrix is such that $A(v)w_i$ has the maximum entry c_{max}^i in row i, and is the same constant $c^i < c_{max}^i$ in all other rows. Such vectors exist: If w_s is completely labelled, extend the edge with labels $I - \{i\}$ into the best reply region with label i. Then any point that lies on the extension of the edge in the best reply region with label i has this property. If a label $i \in I$ represents an unplayed strategy, choose the vertex of $X^\triangle$ that represents the unplayed strategy i. In this case, w_i is itself a unit vector such that $A(v)w_i = e_i$. The construction of $w_s^\triangle$ is depicted in Figure 2.10, in which label 1 represents an unplayed strategy. Then $w_s^\triangle$ is the $(m-1)$-simplex spanned by w_i, $i \in I$.

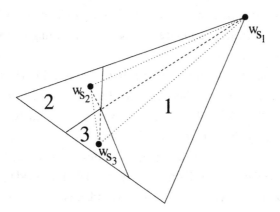

Fig. 2.10. The construction of $w_s^\triangle$

Now label each vertex w_i with label i. This means that $w_s^\triangle$ is an $(m-1)$-simplex whose vertices are completely labelled, i.e. have all labels $i \in I$. This induces an ordering of the vertices of $w_s^\triangle$. The simplex $X^\triangle$ is also an $(m-1)$-simplex that is completely labelled, spanned by the vertices $-m\hat{v}e_i$ with label

$i, i \in I$. To define the orientation in $X_*^\triangle$, choose the orientation of $X^\triangle$ as the standard orientation. The expression (1.7) for the vertices of $X^\triangle$ is given by $(-1)^m$. Let the coordinates of w_i with respect to the unit vectors be given by w_i^u. So, if $v_1, \ldots, v_m$ are the vertices of $v^\triangle$, described as column vectors with respect to the unit vectors, then $w_i^u = [v_1, \ldots, v_m]w_i$. Then the index of an equilibrium is defined as follows.

Definition 2.9. *The index of an equilibrium represented by $w_s \in X_*^\triangle$ is $+1$ if $w_s^\triangle$ lies in the same orientation class as $X^\triangle$, and it is -1 otherwise. That is, the index is defined as*

$$\text{sign } (-1)^m \det[w_1^u, \ldots, w_m^u] = \text{sign } (-1)^m \det[v_1, \ldots, v_m][w_1, \ldots, w_m]. \quad (2.12)$$

Proposition 2.10 below shows that the index in Definition 2.9 is the same as that in Definition 1.4. It follows that the index as defined here does not depend on the particular vertices of $w_s^\triangle$ chosen. Furthermore, the index is well-defined and does not depend on whether one uses $X_*^\triangle$ or $Y_*^\triangle$. It also follows that the definition is independent of the labelling of the strategies. This can also be seen as follows. Re-labelling the strategies of player I would induce a re-labelling of regions in $X_*^\triangle$, without affecting them as such. Therefore, a re-labelling of the strategies induces the same re-labelling of the vertices of $X^\triangle$ as of the vertices of $w_s^\triangle$.

An illustration of Definition 2.9 is given in Figure 2.11. The pure strategy equilibrium where player I plays pure strategy 1, represented by w''_s, has index $+1$. The labels around w''_s read $1, 2, 3$ in anti-clockwise direction, and so do the labels of the vertices of $X^\triangle$, which are the corners of $X_*^\triangle$. The labels around w'_s read $1, 3, 2$ in anti-clockwise direction or $1, 2, 3$ in clockwise direction. Hence the index is defined as -1. The labels around w_s are oriented as the labels of the corners of $X_*^\triangle$, hence the index is $+1$.

Thus, as described in Section 1.1, the index can be identified with a permutation of the labels I. In particular, if, for example, strategies $i_1, \ldots, i_k$, are played with zero probability in an equilibrium w_s, then the $(k-1)$-face of $w_s^\triangle$ that is spanned by the vertices of $w_s^\triangle$ representing labels $i_1, \ldots, i_k$ is the same as the $(k-1)$-face of $X^\triangle$ spanned by the outer vertices representing labels $i_1, \ldots, i_k$. Choosing the orientation of $X^\triangle$ as the standard, this implies that the associated permutation of the labels I is the identity on the subset

$\{i_1, \ldots, i_k\}$. It follows that pure strategy equilibria have index $+1$. If (v, w) is a pure strategy equilibrium in which strategy i of player I is played with probability 1, the permutation of the labels I is the identity on the labels $I - \{i\}$. But then it must be the identity on $\{i\}$. So the permutation is the identity and has sign $+1$. This can also be verified using the expression (2.12), noting that the entries of w_i'' are less than zero.

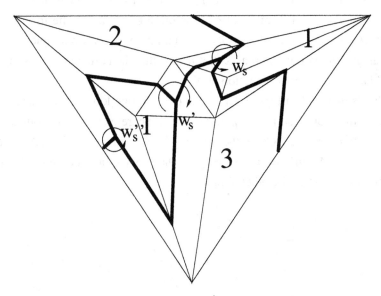

Fig. 2.11. The index in $X_*^\triangle$ for Example 2.3

The above definition of index uses the orientation in $X_*^\triangle$, which is the projection of the labelled surface $P_*^\triangle$. One can also define the orientation by using the labelled surface $P_*^\triangle$ directly. In the same way as the simplex $w_s^\triangle$ is constructed in $X_*^\triangle$, one can construct $w_s^\triangle$ in $P_*^\triangle$ such that it lies on the facet $v^\triangle$ of $P_*^\triangle$ that contains w_s. These simplices are also denoted as $w_s^\triangle$.

To define the index in $P_*^\triangle$, one has to account for the fact that the projection has an effect on the orientation of simplices. Let $w_s^\triangle$ be a simplex around an equilibrium w_s contained in $v^\triangle$, where $v^\triangle$ is a facet of $P^\triangle - X^\triangle$. Then the sign in (1.7) for the vertices of $w_s^\triangle$, ordered by their labels, is the opposite as the sign in (1.7) for the vertices of the projected simplex.

To see this, note that the expression (1.7) for vertices of a simplex on $P_*^\triangle -$ $X^\triangle$ is the same as (1.8) for vertices of the simplex relative to the projection point $v_p = (-m\hat{v}, \ldots, -m\hat{v})$. This is due to the fact that both points $\mathbf{0} \in \mathbb{R}^m$ and $v_p = (-m\hat{v}, \ldots, -m\hat{v})$ lie in the same of the two halfspaces which are defined by the hyperplane containing the simplex. Furthermore, the expression (1.8) for a simplex $w_s^\triangle$ relative to v_p is not affected by the projection of $w_s^\triangle$ on $X^\triangle$. For the simplex $X^\triangle$, the expression (1.7) for the ordered vertices of $X^\triangle$ is the negative as that in (1.8) relative to v_p . Both $\mathbf{0} \in \mathbb{R}^m$ and v_p lie in different halfspaces defined by the hyperplane containing $X^\triangle$. So if a simplex $w_s^\triangle$ in $X_*^\triangle$ has the same orientation as $X^\triangle$, it means that the corresponding simplex in $P_*^\triangle$ has the opposite orientation as $X^\triangle$.

This is depicted in Figure 2.12. One the left, one looks at the surface of $P^\triangle$ from the projection point v_p through $X^\triangle$, where v_p lies on the outside of $P^\triangle$. On the right, one looks at the surface of $P^\triangle$ from $\mathbf{0} \in \mathbb{R}^m$, which lies the inside of $P^\triangle$. Moving from v_p to $\mathbf{0} \in \mathbb{R}^m$ changes the orientation of $X^\triangle$, but not the orientation of the other simplices.

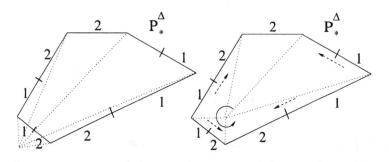

Fig. 2.12. The index in $P_*^\triangle$

Hence, in $P_*^\triangle$ the index of an equilibrium w_s is $+1$ if $w_s^\triangle$ has the opposite orientation as $X^\triangle$, and it has index -1 otherwise. This means that the artificial equilibrium itself has, by definition, index -1. So let, as before, $w_1, \ldots, w_m$ be the set of vertices of $w_s^\triangle$ described by their coordinates with respect to the vertices of $v^\triangle$, where $v^\triangle$ is a facet of $P^\triangle$. Let the vertices of $v^\triangle$ be given as $v_1, \ldots, v_m$, described as column vectors with respect to the unit vectors as basis. Let $w_1^u, \ldots, w_m^u$ denote the set of vertices of $w_s^\triangle$ described by their coordinates with respect to the unit vectors as basis. So $w_i^u = [v_1, \ldots, v_m]w_i$. Then the index is given by

$$\text{sign}(-1)^{m+1}\det[w_1^u, \ldots, w_m^u] = \text{sign}(-1)^{m+1}\det[v_1, \ldots, v_m][w_1, \ldots, w_m].$$
(2.13)

So the index as in (2.13) for the construction $P_*^\triangle$ is the negative of the expression (2.12) for the construction $X_*^\triangle$. This accounts for the effect of the projection on the orientation.

Proposition 2.10. *The index as in Definition 2.9 is the same as the index in Definition 1.4.*

Proof. Without loss of generality, it can be assumed that the entries of the payoff matrices A and B are strictly greater than zero. Consider the labelled surface $P^\triangle$. Let (v, w) be an equilibrium, and let $w_s^\triangle$ be the corresponding completely labelled simplex contained in the facet $v^\triangle$ of $P^\triangle$. The simplex $v^\triangle$ is spanned by some vectors $v_1, \ldots, v_m$, which are described as column vectors with respect to the unit vectors as a basis. These vectors are some m vertices of the polar polytope $P^\triangle$ as in (2.3).

If v_l represents a strategy j of player II, then $v_l = \lambda_j B_j$, where $\lambda_j = \frac{\hat{v}}{\hat{v} - B_j}$ is a positive scalar (compare (2.2)). If v_l represents an unplayed strategy i of player I, then $v_l = -m\hat{v}e_i$. So $v_l = -\lambda_i e_i$, where $\lambda_i = m\hat{v}$ is a positive scalar.

Let $w_1, \ldots, w_m$ denote the ordered set of vertices of $w_s^\triangle$, given by their coordinates with respect to the vertices of $v^\triangle$. These vectors are, by construction, such that $A(v)w_i$ has the maximum entry c_{max}^i in row i, and is the same constant $c^i < c_{max}^i$ in all other rows. Let C denote the matrix $A(v)[w_1 \ldots w_m]$. Then $\det C$ has positive sign, since any convex combination of C with the identity matrix has full rank. Note that all entries of C are strictly greater than zero, since all entries of A are strictly greater than zero.

One obtains $[w_1, \ldots, w_m] = A(v)^{-1}C$. With respect to the unit vectors, the vertices of $w_s^\triangle$ are given by the vectors $[w_1^u, \ldots, w_m^u] = \tilde{B}[w_1, \ldots, w_m]$, where $\tilde{B} = [v_1, \ldots, v_m]$. The rows of $\tilde{B}$ can be ordered such that if row j of $\tilde{B}$ represents an unplayed strategy, then $\tilde{B}_j = -\lambda_j e_j$. If the rows of $\tilde{B}$ are ordered in this way, then the j-th column of $A(v)$ is given by $A(v)_j = e_j$.

Let k denote the size of the support in (v, w), and let A' and B' be defined as in (1.10). For the expression in (2.13), this gives

$$\text{sign } (-1)^{m+1} \det [w_1^u \ldots w_m^u] = \text{sign } (-1)^{m+1} \det \left[\tilde{B} A(v)^{-1} C \right]$$
$$= \text{sign } (-1)^{k+1} \det B' \det A'. \quad (2.14)$$

Note that sign det $A(v)^{-1} = \text{sign det } A(v) = \text{sign det } A'$, since $A(v)_j = e_j$ if column j represents an unplayed strategy. One also has sign det $C = +1$. Furthermore, sign det $\tilde{B} = (-1)^{m-k}$sign det B'. This is due to the fact that the rows of $\tilde{B}$ are ordered such that if row j of $\tilde{B}$ represents an unplayed strategy, then $\tilde{B}_j = -\lambda_j e_j$ with $\lambda_j > 0$. All other rows of $\tilde{B}$ are positive multiples of columns of B. Thus the expression in (2.13) is the same as the expression in Definition (1.4). $\qquad \square$

The expression in (2.14) can be interpreted as follows. The term $(-1)^{k+1}$ accounts for the alternating sign of the matrix corresponding to $X^\triangle$, sign det B' gives the orientation of $v^\triangle$, and sign det A' gives the orientation of $w_s^\triangle$ within $v^\triangle$.

In $X_*^\triangle$, the artificial equilibrium is not represented as such. Instead, it can be thought of lying under $X_*^\triangle$, since it is covered by the projection of $P_*^\triangle - X^\triangle$. Alternatively, the artificial equilibrium can be represented in $X_*^\triangle$ by attaching a mirrored version of $X^\triangle$ along some $(m-2)$-face to $X_*^\triangle$ as depicted in Figure 2.13. The representation of the index in $X_*^\triangle$ allows to intuitively show that indices which are connected via a L-H path have opposite indices. This result was first proven by Shapley (1974).

Proposition 2.11. *Equilibria connected by an L-H path have opposite indices. The sum of indices of equilibria in a non-degenerate bimatrix game is $+1$.*

Proof. The proof is illustrated in Figure 2.13. Note that the dual construction can also be applied to player II's strategy space Y to obtain $Y_*^\triangle$ to follow L-H paths defined by a missing label $j \in J$. The proof here applies to $X_*^\triangle$ and L-H paths defined by a missing label $k \in I$ of player I. The proof for L-H paths in $Y_*^\triangle$ is equivalent.

Take two equilibria (v_1, w_1) and (v_2, w_2) that are connected in $X \times Y$ via an L-H path in $M(k)$ for some $k \in I$. In $X_*^\triangle$, this corresponds to two completely labelled points w_{s1} and w_{s2} that are completely labelled and are connected in $X_*^\triangle$ by some path in $M_*^\triangle(k)$. Along the path, the relative position of the regions with labels $I - \{k\}$ is constant. Fixing the face with labels $I - \{k\}$, the vertex with label k lies on one side in $w_{s1}^{\triangle}$, and on the other side in $w_{s2}^{\triangle}$, so $w_{s1}^{\triangle}$ and $w_{s2}^{\triangle}$ must have opposite indices (see e.g. Eaves and Scarf (1976) or Garcia and Zangwill (1981, Theorem 3.4.1)).

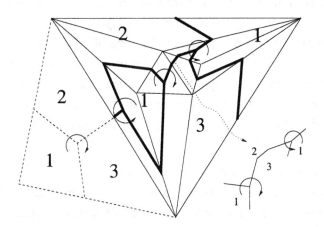

Fig. 2.13. Orientation along L-H paths

As argued above, the artificial equilibrium has orientation -1. Since for a given missing label the L-H paths always yield equilibrium pairs (including the artificial equilibrium), the sum of indices of equilibria equals 0 if one also counts the artificial equilibrium, and it equals $+1$ if one does not.   □

Proposition 2.10 shows that the index is independent of unplayed strategies. This is also illustrated by the dual construction, since the permutation of the labels representing unplayed strategies is trivial. The following observation shows that this invariance property, together with the fact that the sum of indices of equilibria of a game equals $+1$, actually defines the index.

Proposition 2.12. *Let* $Ind(v,w)$ *be some index function that assigns an index* $+1$ *or* -1 *to equilibria* (v,w) *of a non-degenerate bimatrix game. If* $Ind(v,w)$ *is such that the indices of equilibria of a game add up to* $+1$ *and such that the index does not depend on unplayed strategies, then* $Ind(v,w)$ *must be the same as in Definition 1.4.*

Proof. The proof is by induction on the number k of strategies played in equilibrium. The case $k = 1$ reflects pure strategy equilibria, for which both concepts yield index $+1$. Now fix a non-degenerate bimatrix game G, and consider an equilibrium of G in which each player plays k strategies. Consider the game $k \times k$ bimatrix game G' that is obtained from the original game G by deleting all unplayed strategies, i.e. consider the game with payoff matrices A' and B'. Then the equilibrium is the only completely mixed equilibrium in G'. The sum of indices of the equilibria of G' equals $+1$ with respect to both $Ind(\cdot)$ and Definition 1.4. But for all equilibria of G' that use $k-1$ or less strategies, both indices are the same, noting that both concepts only depend on the strategies played in equilibrium. The sum of indices of the equilibria of G' equals $+1$, thus the indices of the completely mixed equilibrium of G' must coincide. These, in turn, are the same as the indices of the equilibrium as an equilibrium of G.   □

In the same way as in the proof of Proposition 2.12, one can show that the invariance property, i.e. the index does not depend on unplayed strategies, and the property that equilibria at the ends of L-H paths have opposite indices completely characterise the index.

3

Sperner's Lemma and Labelling Theorems

This chapter shows how the labelled dual construction $X_*^\triangle$ relates to labelled triangulations as in Sperner's Lemma. Sperner's Lemma is a result from combinatorial topology that applies to triangulations of the unit simplex together with a labelling of the vertices in the triangulation. Sperner's Lemma states the existence of a fully labelled simplex if a certain boundary condition is satisfied. This condition is a restriction on the labelling function for vertices on the boundary.

Sperner's Lemma is equivalent to Brouwer's fixed point theorem (see e.g. Garcia and Zangwill (1981)). Since the Nash equilibria of a game can be described as the fixed points of a suitable mapping $f : X \times Y \to X \times Y$, a "connection" between Sperner's Lemma and bimatrix games is nothing new. What is new, however, is the fact that the dual construction for $m \times n$ bimatrix games relates to Sperner's Lemma in dimension $m - 1$. This also allows one to show that the existence of a Nash equilibrium in an non-degenerate $m \times n$ bimatrix game implies Brouwer's fixed point theorem in dimension $m - 1$. Since Nash equilibria can, conversely, be described as fixed points, Brouwer's fixed point theorem is equivalent to the existence of Nash equilibria in non-degenerate bimatrix games.

The structure of this chapter is as follows. Section 3.1 reviews Sperner's Lemma in its classical form. It shown that Sperner's Lemma is equivalent to the KKM Lemma, a classical result by Knaster, Kuratowski and Mazurkiewicz (1929), and to Brouwer's fixed point theorem. In Section 3.2 it is shown how these results apply to bimatrix games. In particular, it is shown that for every labelled regular triangulation $|\triangle^{m-1}|$ with no vertices on the boundary other than the unit vectors e_i with label i, there exists an $m \times n$ non-degenerate bimatrix game such that the labelled dual construction for the game is equivalent to the labelled triangulation (Proposition 3.9). The L-H algorithm in that bimatrix game is equivalent to a well-known algorithm that finds completely labelled simplices. It is also shown that for every labelled dual construction $X_*^{\triangle}$ there exists a refinement of $|X^{\triangle}|$ and a labelling of the vertices that is consistent with the best reply regions such that the Nash equilibria are represented by the completely labelled simplices (Proposition 3.14). The relation of the dual construction to Sperner's Lemma is then used to show that the existence of Nash equilibria in non-degenerate bimatrix games is equivalent to Brouwer's fixed point theorem (Corollary 3.13). Section 3.3 translates the division of $X_*^{\triangle}$ into a mapping that characterises the Nash equilibria. This section is important, as it lies the technical foundation of the subsequent chapters.

3.1 Sperner's Lemma

Sperner's Lemma (Sperner (1928)) applies to triangulations of a simplex with labelled vertices. Sperner's lemma states that there exists an odd number of completely labelled simplices in a labelled triangulation of the standard $(m - 1)$-simplex $\triangle^{m-1}$ if a boundary condition is fulfilled. This boundary condition states that the label of a vertex v on the boundary is one of the labels of the vertices that span the face that contains v. Sperner's Lemma is a classical result from combinatorial topology and is equivalent to Brouwer's fixed point theorem and the KKM Lemma (see e.g. Garcia and Zangwill (1981)).

A *triangulation* (or *simplicial subdivision*) of $\triangle^{m-1}$, denoted as $|\triangle^{m-1}|$, is a finite collection of smaller $(m-1)$-simplices whose union is the simplex, and that is such that any two of the simplices intersect in a face common to both, or the intersection is empty. Let V denote the set of vertices of the smaller simplices in $|\triangle^{m-1}|$. A labelling function is a function that assigns a label $i \in I = \{1, \ldots, m\}$ to each vertex $v \in V$, i.e. $L : V \to I$. An example of a triangulation of $|\triangle^{m-1}|$ with a labelling L is depicted in Figure 3.1. A triangulation together with a labelling of the vertices is referred to as *labelled triangulation*.

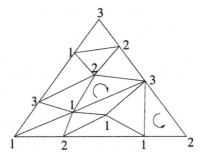

Fig. 3.1. A labelled triangulation

The simplex $\triangle^{m-1}$ is spanned by the unit vectors $e_i \in \mathbb{R}^m$, $i \in I$, where $I = \{1, \ldots, m\}$. The Sperner boundary condition, which is referred to as the *Sperner condition*, states that if a vertex $v \in V$ lies on the $(k-1)$-face of $\triangle^{m-1}$ that is spanned by e_j, $j \in I_k$, with $I_k = \{i_1, \ldots, i_k\} \subset I$, then $L(v) \in I_k$. Note that the Sperner condition only restricts the labelling of vertices that lie on the boundary ($I_k \subset I$ and $I_k \neq I$). For vertices in the interior of $\triangle^{m-1}$ there is no restriction ($I_k = I$). So it is appropriate to refer to the Sperner condition as a boundary condition. The Sperner condition implies that the unit vectors e_i have label i. So every vertex v can only be assigned one of the labels of those vertices that span the (minimal) face that contains v. For the example in Figure 3.1, the Sperner condition is fulfilled. For example, the vertices that lie on the boundary face spanned by vertices with labels 1 and 2 only have labels 1 or 2.

Definition 3.1 (Sperner condition). *Let $v \in V$ be contained in a $(k-1)$-face of $\triangle^{m-1}$ spanned by e_j, $j \in I_k$, with $I_k = \{i_1, \ldots, i_k\} \subset I$, and let k be minimal in this respect. Then a labelling $L: V \to I$ fulfils the Sperner condition if $L(v) \in I_k$.*

Sperner's Lemma states that there exists an odd number of completely labelled simplices if the Sperner condition is satisfied. A simplex is called completely labelled if the vertices of the simplex have distinct labels, i.e. if the vertices have labels $1, \ldots, m$. It follows that there exists at least one completely labelled simplex. Sperner's Lemma also states that there exists one more completely labelled simplex with positive orientation than with negative orientation. An orientation is an equivalence class as described through

(1.7). According to (1.7), the sign of the determinant associated with the unit simplex $\triangle^{m-1}$ with vertices labelled $L(e_i) = i$ is $+1$. If a simplex is completely labelled, one can order the vertices according to their labelling. Applying (1.7) and choosing the orientation of the unit simplex as the standard orientation, one can define the orientation of a completely labelled simplex.

Definition 3.2 (Orientation). *A completely labelled simplex has orientation $+1$, if it falls in the same equivalence class as the unit simplex $\triangle^{m-1}$ with vertices labelled $L(e_i) = i$, and -1 otherwise.*

The labels of a completely labelled simplex can be seen as an ordering of its vertices, and the orientation of a fully labelled simplex corresponds to a permutation of the labels of the vertices as described before. The orientation is $+1$ if the permutation has sign $+1$, and it is -1 otherwise. For the example in Figure 3.1, the completely labelled simplex in the bottom right corner has orientation $+1$; the labelling reads $(1,2,3)$ in anti-clockwise direction. The completely labelled simplex in the centre of Figure 3.1 has orientation -1; its labelling reads $(1,2,3)$ in clockwise direction.

Theorem 3.3 (Sperner's Lemma). *Consider a labelled triangulation $|\triangle^{m-1}|$ such that the labelling satisfies the Sperner condition. Then there exists an odd number of completely labelled simplices, one more with orientation $+1$ than with orientation -1.*

Proof. This proof employs methods from combinatorial topology and is by induction (see e.g. Henle (1994, p. 38) for the case $m = 3$). The case for $m = 1$ is trivial, and $m = 2$ is also easy to verify. So suppose the claim is true for triangulations of $\triangle^{m-2}$.

Fix a label $k \in I$, and consider a simplex $\triangle \in |\triangle^{m-1}|$ that is spanned by vertices $v_1, \ldots, v_m$. Consider an $(m-2)$-face of $\triangle$ that is spanned by, say, vertices $v_1, \ldots, v_{m-1}$. Relative to $\triangle$, each $(m-2)$-face has an orientation induced by the orientation of $\triangle^{m-1}$ and the labels $I - \{k\}$: If the $m-1$ vertices of the face do not have labels $I - \{k\}$, the orientation is 0. If the vertices of the face have $m-1$ distinct labels $I - \{k\}$, then the orientation of the $(m-2)$-face is the orientation of the completely labelled simplex that would be obtained by giving v_m the missing label k. This is depicted in Figure 3.2 for $k = 1$. There are three cases.

1) A simplex $\triangle$ does not have labels $I - \{k\}$. In this case the orientations of its $(m-2)$-faces are zero since no $(m-2)$-face can have labels $I - \{k\}$. Hence the sum of the orientations over the $(m-2)$-faces of $\triangle$ is zero.

2) A simplex $\triangle$ has exactly the $m-1$ distinct labels $I - \{k\}$. Then exactly two $(m-2)$-faces of $\triangle$ are such that they have the same $m-1$ distinct labels $I - \{k\}$, while all other $(m-2)$-faces have labels other than $I - \{k\}$. The latter ones have by definition orientation zero, while the two former ones are such that they have opposite orientations. Hence the sum of orientations over the $(m-2)$-faces of $\triangle$ is also zero.

3) A simplex $\triangle$ is completely labelled. Then, by definition, their exists exactly one $(m-2)$-face of $\triangle$ with labels $I - \{k\}$. This face has orientation $+1$ if $\triangle$ has positive orientation, and orientation -1 if $\triangle$ has negative orientation.

Now consider an $(m-2)$-face that lies in the interior of $\triangle^{m-1}$. By definition, it belongs to exactly two simplices that are adjacent. With respect to one simplex its orientation is the negative of its orientation with respect to the other simplex (including the case where the orientation is zero). So, adding up the orientations of all $(m-2)$-faces of all simplices in $|\triangle^{m-1}|$, this sum must equal the sum of orientations of the boundary $(m-2)$-faces of $|\triangle^{m-1}|$, since the orientations of $(m-2)$-faces in the interior cancel out.

Boundary $(m-2)$-faces of $|\triangle^{m-1}|$ with labels $I - \{k\}$ can only lie on the $(m-2)$-face spanned by e_i, $i \in I - \{k\}$. But the sum of orientations of these $(m-2)$-simplices equals $+1$ by induction assumption. Hence, there exists exactly one more completely labelled simplex with positive orientation than with negative orientation. Note that the proof is independent of the label k chosen for the proof. $\qquad\square$

An illustration of the proof in the case $m = 3$ is depicted in Figure 3.2 for the example in Figure 3.1. Consider a triangle $\triangle \in |\triangle^2|$, and fix the label $k = 1$. The assigned orientation is $+1$ if the edge has labels $2, 3$ oriented in the same way as the edge $2, 3$ in the original simplex, and -1 if it has labels $2, 3$ oriented in the opposite way. All other edges have orientation 0. Now consider two triangles $\triangle$ and $\triangle'$ that share an edge. Then the edge in one triangle has the opposite orientation as the same edge in the adjacent simplex. The sum of orientations of the edges of a triangle is either $+1$, -1 (if completely labelled)

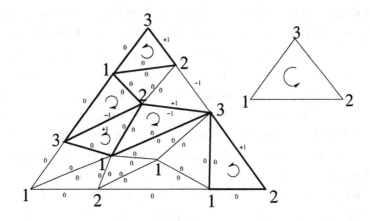

Fig. 3.2. The proof of Sperner's Lemma for $\triangle^2$

or 0 (if not completely labelled). But adding up the sums of orientations of edges over all triangles in $|\triangle^2|$ is the same as the sum of orientations of edges on the boundary of $|\triangle^2|$, since the orientations of edges in the interior of $|\triangle^2|$ cancel out. The Sperner condition ensures that this outer sum is $+1$. Boundary edges with labels $2,3$ can only lie on the $(m-2)$-face of $\triangle^2$ spanned by e_2 and e_3. On this 1-face, the orientations add up to $+1$. Hence, there exists an odd number of completely labelled simplices, one more with positive orientation than with negative orientation. In Figure 3.2 these are depicted by bold edges.

So the Sperner condition, which is a restriction of the labelling on the boundary, determines the existence of a completely labelled simplex. An alternative proof of Theorem 3.3 can be given by using degree theory from algebraic topology, described next. This proves useful when comparing the Sperner situation with the labelled dual construction $X_*^{\triangle}$ and when formalising a generalised version of Sperner's Lemma that applies to components of equilibria in Chapter 5. For this, one translates the labelled triangulation into a mapping between two standard $(m-1)$-simplices. The mapping also yields a division of $\triangle^{m-1}$ into labelled regions such that one can apply the KKM Lemma (see below).

Definition 3.4. *Consider the standard $(m-1)$-simplex $\triangle^{m-1}$. Then $\triangle^{m-1}$ is the (non-disjoint) union of m convex regions $\triangle^{m-1}(i)$ with labels $i \in I$ as follows: $\triangle^{m-1}(i) = \{x \in \triangle^{m-1} \mid x_i = \max_{k\in I} x_k\}$. This division of $\triangle^{m-1}$ into convex regions is referred to as the canonical division and is denoted as $\triangle_*^{m-1}$. Each point in $p \in \triangle_*^{m-1}$ is assigned the labels of the regions that contain p, i.e. $L(p) = \{i \in I \mid p \in \triangle^{m-1}(i)\}$. The vertices of $\triangle_*^{m-1}$ are the vertices of the sets $\triangle^{m-1}(i)$, $i \in I$. The completely labelled point in the centre of $\triangle_*^{m-1}$ is denoted as v_*.*

Essentially, the division of $\triangle_*^{m-1}$ into labelled regions is same as the division of $X = \triangle^{m-1}$ into best reply regions in the $m \times m$ coordination game with identity matrices as payoffs, and the vertices of $\triangle_*^{m-1}$ are the vertices in $X = \triangle^{m-1}$. A depiction of the canonical division is given in Figure 3.3.

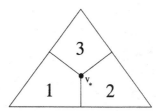

Fig. 3.3. The canonical division $\triangle_*^{m-1}$

The labelling now defines a mapping f^S from $|\triangle^{m-1}|$ to $\triangle_*^{m-1}$. Consider a simplex $\triangle \in |\triangle^{m-1}|$ that is spanned by vertices $v_1,\ldots,v_m$. Each vertex has a label $L(v_i)$, and is mapped to the vertex $e_{L(v_i)}$ in $\triangle_*^{m-1}$. This mapping preserves the labels of the vertices, i.e. $L(v) = L(f^S(v))$. Having defined the mapping on the vertices of $\triangle$, it can be linearly extended to a mapping from $\triangle$ by mapping a convex combination of vertices on the convex combination of their images, i.e. $f^S(\sum_{i=1}^m \lambda_i v_i) = \sum_{i=1}^m \lambda_i f^S(v_i)$.

It is easy to verify that f^S maps every k-face of a simplex in $|\triangle^{m-1}|$ on some k-face of $\triangle_*^{m-1}$. In particular, if the $k+1$ vertices of a k-face have distinct labels $i_1,\ldots,i_{k+1}$, it is mapped affinely on the k-face of $\triangle_*^{m-1}$ that is spanned by unit vectors $e_{i_1},\ldots,e_{i_{k+1}}$. If the $k+1$ vertices of that face have labels $i_1,\ldots,i_l$ (with $l \leq k+1$, so some labels might be duplicate), it is mapped on the $(l-1)$-face of $\triangle_*^{m-1}$ that is spanned by unit vectors $e_{i_1},\ldots,e_{i_l}$. Since this also holds for the $(m-2)$-faces that lie on the boundary of $|\triangle^{m-1}|$, the mapping f^S maps boundary on boundary, i.e.

$$f^S : (|\triangle^{m-1}|, \partial |\triangle^{m-1}|) \longrightarrow (\triangle^{m-1}_*, \partial\triangle^{m-1}_*). \tag{3.1}$$

The mapping in (3.1) is referred to as the *Sperner mapping*, and induces a division of $|\triangle^{m-1}|$ into labelled regions $|\triangle^{m-1}|(i)$. This is depicted in Figure 3.4. These regions are the pre-images of the regions $\triangle^{m-1}(i)$ in the canonical division $\triangle^{m-1}_*$. This division of $\triangle^{m-1}$ into labelled regions is denoted as $|\triangle^{m-1}|_*$. The subscript "*" symbolizes a division into labelled regions (as in the case $X^\triangle_*$). The labels of a point $p \in |\triangle^{m-1}|_*$ are defined as $L(p) = L(f(p))$. The bold numbers and lines in Figure 3.4 mark the regions $|\triangle^{m-1}|(i)$. In this representation, the completely labelled points correspond to completely labelled simplices, since only the centre of completely labelled simplices is mapped to v_*.

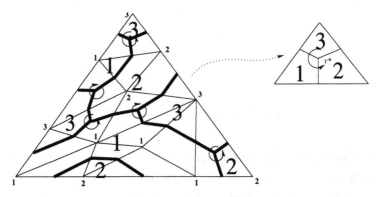

Fig. 3.4. A division of $\triangle^{m-1}$ into labelled regions

Alternatively, let $v_1, \ldots, v_m$ be the vertices of some simplex $\triangle$ in $|\triangle^{m-1}|$ with labels $L(v_i)$, for $i \in I$. A point in $\triangle$ is given by its coordinates p with respect to $v_1, \ldots, v_m$. Then, on each $\triangle$, the mapping f^S can be described by the matrix $A^S(\triangle) = [e_{L(v_1)} \cdots e_{L(v_m)}]$. This matrix is referred to as the *Sperner matrix*. So a point in $\triangle$ with coordinates p is mapped to $A^S(\triangle)p$. The labels of a point with coordinates p are given by $L(p) = \{k \in I \mid (A^S(\triangle)p)_k = max_{i \in I}(A^S(\triangle)p)_i\}$. So the division into labelled regions is obtained in a similar way as the labelled dual construction is obtained via $A(v)$. The difference is that in the Sperner case the columns of the matrix $A^S(\triangle)$ are unit vectors, whereas in case of $A(v)$ the matrix consists of a mixture of payoff vectors and unit vectors.

The Sperner condition determines the degree of the Sperner mapping f^S. The concept of degree is a useful tool that incorporates what was done "manually" in the proof of Theorem 3.3. For the mapping f^S, the degree counts the number of pre-images of the completely labelled point $v_* \in \triangle_*^{m-1}$, where each pre-image is counted with its local degree. The local degree at a pre-image of v_* equals the orientation of the completely labelled simplex that contains the pre-image. For a mapping that permutes the vertices of a simplex, the degree equals the sign of the permutation. In Figure 3.4, this is depicted by the oriented arc around completely labelled points.

Furthermore, the degree of a mapping is the same as the degree of the mapping restricted to the boundary. The degree of f^S restricted to the boundary of $\triangle^{m-1}$ counts, for an arbitrary but fixed label $k \in I$, the number of almost completely labelled points on the boundary $|\triangle^{m-1}|_*$ with labels $I - \{k\}$, again counting each with its local degree. The local degree of f^S restricted to the boundary equals the orientation that was assigned to $(m-2)$-faces in the proof of Theorem 3.3. In particular, it is independent of the label k chosen.

The two paragraphs above contain all that is needed in terms of degree theory for the remainder of this work. A detailed account of the degree can e.g. be found in Dold (1972, IV, 4 and 5).

Lemma 3.5. *If the Sperner condition is satisfied then the degree of the Sperner mapping f^S is $+1$.*

Proof. The proof is by induction. For $m = 1$ the case is trivial (and for $m = 2$ it is also easy to check). So suppose the statement is true for triangulations of the standard $(m-2)$-simplex. Fix a label $k \in I$. In the division of $\triangle_*^{m-1}$ into labelled regions consider the vertex v with labels $I - \{k\}$ that lies on the $(m-2)$-face spanned by unit vectors e_i, $i \in I - \{k\}$. Now restrict f^S to the boundary. For f^S restricted to the boundary, the pre-images of v can only lie on the $(m-2)$-face of $|\triangle^{m-1}|$ that is spanned by e_i, $i \in I - \{k\}$ (see also Figure 3.4). This is ensured by the Sperner condition. But then the degree of f^S restricted to the boundary is $+1$ by induction assumption, which equals the degree of f^S. $\square$

After translating the labelling into a mapping, Sperner's Lemma is simply a consequence of Lemma 3.5. The degree of f^S equals $+1$. This degree is, as explained above, the sum of local degrees at pre-images of v_*. But the local degree at a pre-image of v_* is the same as the orientation of the completely labelled simplex that contains the pre-image.

The induced division $|\triangle^{m-1}|_*$ is a division to which one can apply the KKM Lemma, a classical result by Knaster, Kuratowski and Mazurkiewicz (1929).

Theorem 3.6 (KKM Lemma). *Let C_i, with $i \in I = \{1,\dots,m\}$, be a collection of closed subsets of $\triangle^{m-1}$ such that for all subsets $I_k \subset I$ the face of $\triangle^{m-1}$ that is spanned by e_i, for $i \in I_k$, is contained in $\bigcup_{i \in I_k} C_i$. Then $\bigcap_{i \in I} C_i \neq \emptyset$.*

Proof. The KKM Lemma is implied by Sperner's Lemma. To see this assume that $\bigcap_{i \in I} C_i = \emptyset$. Now each subset C_i is closed by assumption, and since it is bounded, it is compact. So the set $\Pi_{i \in I} C_i$ is compact, and the mapping $\Pi_{i \in I} C_i \to \mathbb{R}$ defined by $(x_1,\dots,x_l) \mapsto \max_{i,j} \|x_i - x_j\|$ takes a minimum $\varepsilon > 0$. Therefore there exists an $\varepsilon > 0$ such that for all $x \in \triangle^{m-1}$ the ε-neighbourhood $U_\varepsilon(x)$ around x is such that $U_\varepsilon(x) \cap C_i = \emptyset$ for at least one set C_i. Now choose a triangulation of $\triangle^{m-1}$ such that each simplex in the triangulation has a diameter smaller than ε. Label the vertices v such that $L(v) \in \{i \mid v \in C_i\}$. Then one has a triangulation of $\triangle^{m-1}$ that fulfils the Sperner condition but does not contain a completely labelled simplex. This violates Sperner's Lemma. $\square$

Conversely, it is easy to see that the KKM Lemma implies Sperner's Lemma. Assuming a triangulation of $\triangle^{m-1}$ that fulfils the Sperner condition but does not contain a completely labelled simplex, one obtains a division of $\triangle^{m-1}$ via the Sperner mapping f^S that satisfies the assumptions of the KKM Lemma but does not contain a completely labelled point. Thus Sperner's Lemma is equivalent to the KKM Lemma (see also e.g. Garcia and Zangwill (1981)).

There exists a well-known algorithm that finds a completely labelled simplex in $|\triangle^{m-1}|$ (or a completely labelled point in $|\triangle^{m-1}|_*$). This algorithm is described below, and is referred to as the *Sperner algorithm*. First, "extend" $|\triangle^{m-1}|$ by inscribing it into a larger $(m-1)$-simplex $|\triangle^{m-1}|^e$ as shown in Figure 3.5 (see e.g. Scarf (1983)). This gives a triangulation of the extended simplex that coincides with the triangulation $|\triangle^{m-1}|$ in the interior. Now label the vertices that span $|\triangle^{m-1}|^e$ such that there are no completely labelled simplices except from those in $|\triangle^{m-1}|$. This is possible due to the Sperner condition: Take the outer vertex of the extended structure that lies on the outside of the face of $|\triangle^{m-1}|$ on which the vertices can only have labels $i \in I - \{k\}$. Labelling the outer vertex with $k+1$ (mod m) ensures that no new completely labelled simplices are created. Furthermore, it ensures that, for every set of labels $I - \{k\}$, there exists exactly one $(m-2)$-face on the boundary of $|\triangle^{m-1}|^e$ that has labels $I - \{k\}$.

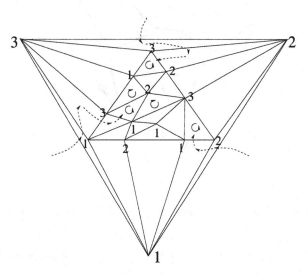

Fig. 3.5. An algorithm for finding completely labelled triangles

The algorithm can now be described as follows (see Figure 3.5). Start from the outside of the extended construction (or at a completely labelled simplex once one has been found). Choose a label $k \in I$ and flip over the $(m-2)$-face that has labels $I - \{k\}$. If the new simplex is not completely labelled, it must have exactly one other $(m-2)$-face (other than the face one flipped over) with the same labels $I - \{k\}$. Then flip over this $(m-2)$-face into an adjacent simplex, and so on. Eventually, this algorithm yields a completely labelled simplex in $|\triangle^{m-1}|$ (see e.g. Scarf (1983)). Simplices that are connected through the algorithm have opposite orientation.

The Sperner algorithm translates easily into the topological setting. Let f^S denote the Sperner mapping from the enlarged simplex $|\triangle^{m-1}|^e$ to $\triangle_*^{m-1}$.

This yields a division of the extended simplex into labelled regions in which the completely labelled simplices correspond to points that are mapped to v_* under f^S. For every label k, there exists exactly one point on the boundary with labels $I - \{k\}$. The path with labels $I - \{k\}$ that starts on the boundary leads to a completely labelled point.

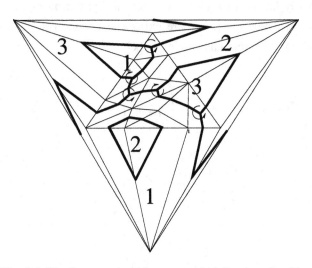

Fig. 3.6. The Sperner algorithm as a path-following algorithm

To emphasise the relevance of Sperner's Lemma in fixed point theory, this section concludes by proving the familiar theorems that show that Sperner's Lemma implies Brouwer's fixed point theorem and vice versa. This also allows one to show in the next section that the existence of Nash equilibria in non-degenerate bimatrix games is equivalent to Brouwer's fixed point theorem.

Theorem 3.7 (Brouwer's fixed point theorem). *Every mapping* $f: \triangle^{m-1} \to \triangle^{m-1}$ *has a fixed point, i.e.* $\exists\, x^* \in \triangle^{m-1} : f(x^*) = x^*$.

Proof. Assume the contrary, i.e. for all $x \in \triangle^{m-1}$ one has $f(x) \neq x$. This defines a mapping $r: \triangle^{m-1} \to \partial\triangle^{m-1}$ that retracts $\triangle^{m-1}$ on its boundary. Define $r(x)$ as the point on the boundary that is given by the intersection point between the line defined by x and $f(x)$ in direction of x and the boundary (see the left picture in Figure 3.7). Since r is continuous and defined on a compact set, the mapping r is uniformly continuous. Now take a triangulation of $\triangle^{m-1}$

into sufficiently small simplices, say with diameter smaller than some δ. Then label the vertices according to $L(v) = L(r(v))$, where $L(r(v))$ is the label of the point $r(v)$ in the canonical division. Then one has a labelling that satisfies the Sperner condition (since r is the identity on the boundary) and is such that no simplex is fully labelled if δ is sufficiently small: Every δ-neighbourhood of x is mapped on some small ε-neighbourhood of $r(x)$, which does not contain more than $m-1$ distinct labels for small ε. This contradicts Sperner's Lemma.

$\square$

Brouwer's fixed point theorem depends on the fact that $\triangle^{m-1}$ cannot be retracted to its boundary. If there exists a subdivision $|\triangle^{m-1}|$ with a labelling that satisfies the Sperner condition and does not contain a completely labelled simplex then the Sperner mapping f^S is a mapping that retracts $\triangle^{m-1}$ to its boundary. Assuming without loss of generality there are no vertices except those of $\triangle^{m-1}$ on the boundary (by inscribing $|\triangle^{m-1}|$ into an extended structure as above), the mapping f^S is the identity on the boundary. Thus the "no-retraction" property implies Sperner's Lemma. But Sperner's Lemma can also be deduced directly from Brouwer's fixed point theorem.

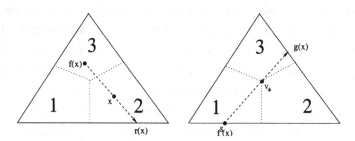

Fig. 3.7. Sperner's Lemma implies Brouwer and vice versa

Proposition 3.8. *Brouwer's fixed point theorem implies Sperner's Lemma and Sperner's Lemma implies Brouwer's fixed point theorem.*

Proof. The latter implication was shown in the proof of Theorem 3.7. So it remains to show that Brouwer's fixed point theorem implies Sperner's Lemma. Suppose one has a labelling that satisfies the Sperner condition and that does not contain a fully labelled simplex. Then the Sperner mapping f^S is such that $f^S(x) \neq v_*$ for all $x \in \triangle^{m-1}$. Then define $g(x)$ as the point on the boundary that is defined as the intersection of the line between $f^S(x)$ and v_* in direction of v_* with the boundary (see the right picture in Figure 3.7). Then $g(x)$ is a mapping for which $g(x) \neq x$ for x in the interior of $\triangle^{m-1}$. Now suppose x lies on some k-face of $\triangle^{m-1}$. By construction of the Sperner mapping, the point $f^S(x)$ lies on that k-face, and the line connecting $f^S(x)$ and v_* does not go elsewhere through this face. So $g(x) \neq x$ for all points on the boundary, and hence g has no fixed points. This contradicts Brouwer's fixed point theorem. $\qquad\square$

3.2 The Application to Bimatrix Games

The division $|\triangle^{m-1}|_*$ into labelled regions induced by the labelled triangulation already shows strong similarities with the labelled dual construction $X_*^{\triangle}$. The division of $|\triangle^{m-1}|_*$ is induced by the Sperner matrix $A^S(\triangle)$ as described on page 66, whereas the division of $X_*^{\triangle}$ is induced by the artificial payoff matrix $A(v)$. The difference, however, is that $A^S(\triangle)$ only consists of unit vectors, whereas $A(v)$ consists of a mixture of unit vectors representing unplayed strategies and columns of A representing pure strategies of player II. So the division of a simplex in $|X^{\triangle}|$ into best reply regions is in general more complex than the division of simplices in $|\triangle^{m-1}|$. Furthermore, the triangulation $|X^{\triangle}|$ is regular as it arises from the projection of a simplicial polytope. The triangulation in the Sperner case can be any triangulation.

Despite the differences, there are still striking similarities between $|\triangle^{m-1}|_*$ and $|X^{\triangle}|_*$, and this section shows how and under what circumstances one can translate one situation into the other and vice versa. The equivalence of Brouwer's fixed point theorem and the existence of Nash equilibria in non-degenerate bimatrix games (Corollary 3.13 below) also shows that these differences are not very deep.

Proposition 3.9. *Let* $|\triangle^{m-1}|$ *be a labelled triangulation of the unit simplex with no vertices on the boundary other than* e_i, *for* $i \in I$. *Let the Sperner condition be satisfied, so* $L(e_i) = i$. *If the triangulation of* $\triangle^{m-1}$ *is regular, then there exists a non-degenerate* $m \times n$ *bimatrix game such that* $|\triangle^{m-1}| = |X^{\triangle}|$ *and* $|\triangle^{m-1}|_* = X_*^{\triangle}$ *(after identifying* $X^{\triangle}$ *with* $\triangle^{m-1}$*).*

Proof. Let $|\triangle^{m-1}|$ be a regular triangulation. Consider the simplex $X^{\triangle}$ that is spanned by the vertices $-m\hat{v}e_i$, for $i \in I$ and some positive constant $\hat{v}$. Then $\triangle^{m-1}$ can be identified with $X^{\triangle}$ via a linear mapping defined by $e_i \mapsto -m\hat{v}e_i$. This mapping induces a regular triangulation $|X^{\triangle}|$ of $X^{\triangle}$. The label of a vertex $v \in |X^{\triangle}|$ is defined by the label of its pre-images.

This yields a labelled and regular triangulation of $X^{\triangle}$. Since the triangulation is regular, the triangulation is the projection of some simplicial polytope $P^{\triangle}$ as in 2.3, with the first m vertices of $P^{\triangle}$ given by $-m\hat{v}e_i$, $i \in I$. The vertices of $P^{\triangle}$ satisfy the conditions in Lemma 2.2 since the triangulation is regular. Also, it can be assumed that $\mathbf{0} \in \mathbb{R}^m$ lies in the interior of $P^{\triangle}$. If not, one could just move the vertices except for $-m\hat{v}e_i$, $i \in I$, along the projection lines to obtain a combinatorially equivalent polytope that contains $\mathbf{0} \in \mathbb{R}^m$. As described in Lemma 2.2, this yields the columns of a payoff matrix B such that the best reply polytope P that arises from B is the polar of $P^{\triangle}$. This determines the payoffs for player II. Note that if there are n vertices in the interior of $|\triangle^{m-1}|$, then the resulting game is of dimension $m \times n$.

Finally, one has to determine the payoff matrix A for player I. These payoffs are determined by the labelling of the vertices. Each vertex $v \in |X^{\triangle}|$ represents a pure strategy of player II. If the label of a vertex is i for some $i \in I$, then define the payoff for player I with respect to the pure strategy that is represented by vertex v as e_i, the unit vector with entry in row i. Then the induced polyhedral division into best reply regions of the simplices in $|X^{\triangle}|$ is the same as the division induced by the labelling of the vertices in $|\triangle^{m-1}|$. The payoff matrix B that induces $|X^{\triangle}|$ is generic. So is the payoff matrix A that only consists of unit vectors and induces the division into best reply regions.   □

Corollary 3.10. *For a missing label $k \in I$ of player I, the L-H algorithm for the game constructed in Proposition 3.9 follows the same path of simplices as the Sperner algorithm.*

Proof. This is an immediate consequence from the construction. The L-H algorithm follows the path of almost completely labelled points in the labelled dual construction. This corresponds to flipping over $(m-2)$-faces in the triangulation which have $m-1$ distinct labels. The labelled dual construction is identical with the division of $\triangle^{m-1}$ that is induced by the Sperner mapping f^S. But the Sperner algorithm also flips over those $(m-2)$-faces in the triangulation that have $m-1$ distinct labels. Hence the paths of both algorithms are identical.   □

Proposition 3.9 is used to conclude Brouwer's fixed point theorem from the existence of Nash equilibria in bimatrix games. The idea of the proof is based on translating a division $|\triangle^{m-1}|_*$ that arises from a Sperner labelling into a division $X_*^{\triangle}$ with a triangulation $|X^{\triangle}|$ that is regular and arises from a payoff matrix B.

For this, consider some triangulation of $\triangle^{m-1}$. Then add a vertex v. Suppose this vertex is contained in some simplex $\triangle$ that is spanned by vertices $v_1, \ldots v_m$. Note that it is allowed for v to lie on some k-face of $\triangle$. Then consider the refinement of $\triangle$ that is given by the simplices spanned by

$$\{v, v_2, \ldots, v_m\}; \ \{v_1, v, v_3, \ldots, v_m\}; \ \ldots; \ \{v_1, \ldots, v_{m-1}, v\}. \tag{3.2}$$

If v lies on the k-face of two or more simplices, the refinement in (3.2) applies to each simplex that contains v. An illustration for this is given on the left in Figure 3.8. First the vertex v is added, then the vertex v', and finally the vertex v''. Note that some of the simplices in (3.2) are not full-dimensional in case v lies on some k-face of $\triangle$ with $k \leq (m-2)$. In this case, they become faces of simplices in the triangulation.

A refinement of a given triangulation that is achieved by iteratively adding vertices at a time to the triangulation is referred to as an *iterated refinement*. The following lemma shows an iterated refinement can divide a simplex into arbitrarily small simplices. The *mesh* of a triangulation is defined as the maximum diameter of a simplex in the triangulation.

Lemma 3.11. *For every* $\varepsilon > 0$ *there exists an iterated refinement of* $\triangle^{m-1}$ *such that the mesh size of the triangulation is smaller than* ε.

Proof. It is shown that the barycentric subdivision is an iterated refinement. The barycentric subdivision is known to produce simplices of arbitrarily small maximal diameter (see e.g. Dold (1972, III, 6)).

A depiction of the barycentric subdivision is given on the right in Figure 3.8. Take a simplex in the triangulation. Then add the barycentre of the

$(m-1)$-simplex as a vertex. Next, add the barycentres of its $(m-2)$-faces as vertices, and continue with the lower dimensional faces and their barycentres. Note that if one adds a vertex to a k-face that is common to more than one simplex in the triangulation, then the vertex is the barycentre of that k-face, i.e. the added vertex is the same for all simplices that contain the k-face. This procedure yields the barycentric subdivision. □

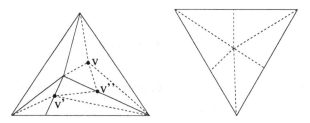

Fig. 3.8. An iterated refinement of a simplex and the barycentric subdivision

Lemma 3.12. *Let* $|X^\triangle|$ *be a regular triangulation of* $X^\triangle$ *with no vertices on the boundary other than those that span* $X^\triangle$. *Then every iterated refinement of* $|X^\triangle|$ *that does not add vertices to the boundary of* $X^\triangle$ *is a regular triangulation. In particular, if* $|X^\triangle|$ *arises from a payoff matrix B, then the refinement arises from an extended payoff matrix that consists of the original columns of B and additional columns.*

Proof. It is required that the added vertices do not lie on the boundary of $X^\triangle$ so that the resulting triangulation can still be achieved as the dual construction for some bimatrix game.

So let $|X^\triangle|$ be a regular triangulation. Then consider the polytope $P^\triangle$ that yields $|X^\triangle|$ via projection. Now take a point v in the interior of $|X^\triangle|$. This point is represented by some point v^P on the boundary of the polytope $P^\triangle$. Now take a point on the line defined by v and v^P that lies outside of $P^\triangle$ but is still close $P^\triangle$. This is depicted in Figure 3.9. Let this point be denoted by c.

Let $P^\triangle$ be defined as the convex hull of points as described in (2.3). Now consider the polytope $P_c^\triangle$ that is given by

$$P_c^\triangle = \text{conv}\{c, c_1, \dots, c_n\}.$$

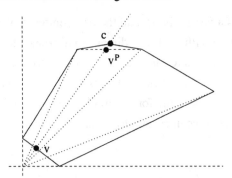

Fig. 3.9. An iterated refinement of $|X^{\triangle}|$

Then c becomes a new vertex of the polytope. Then the vertex c refines the simplicial structure of $P^{\triangle}$ in a way such that the projection of $P_c^{\triangle}$ yields the iterated refinement that is obtained by adding the point v as a vertex. The vertex v is the projection of the vertex c.

For each added point, the polytope $P_c^{\triangle}$ satisfies the requirements of Lemma 2.2. Hence, by Lemma 2.2, one can obtain a payoff matrix that induces the refinement. If the original triangulation arises from a payoff matrix B, the refinement corresponds to a payoff matrix which contains the original columns of B and that has an extra column for each added vertex. $\square$

In Section 3.1 it was shown that Sperner's Lemma is equivalent to Brouwer's fixed point theorem. This section shows how to construct non-degenerate bimatrix games from regular labelled triangulations such that the dual construction has the same properties as the labelled triangulation. Combining these results, one obtains the following result.

Corollary 3.13. *The existence of a Nash equilibrium in a non-degenerate $m \times n$ bimatrix game implies Brouwer's fixed point theorem in dimension $m - 1$. Since Nash equilibria can, conversely, be described as fixed points, Brouwer's fixed point theorem is equivalent to the existence of Nash equilibria in non-degenerate bimatrix games.*

Proof. Consider a mapping $f : \triangle^{m-1} \to \triangle^{m-1}$. Assume $f(x) \neq x$ for all $x \in \triangle^{m-1}$. As in the proof of Theorem 3.7, this yields a retraction r that is defined by the intersection of the line between x and $f(x)$ in direction of x with the boundary of $\triangle^{m-1}$. The mapping r then divides $\triangle^{m-1}$ into labelled regions by considering the pre-images of the labelled regions on $\partial\triangle_*^{m-1}$. In the proof of Theorem 3.7, this division is used to create a labelled triangulation of $\triangle^{m-1}$ such that no simplex is completely labelled. Here, it is shown that one can create a regular labelled triangulation of $\triangle^{m-1}$ with no vertices added to the boundary of $\triangle^{m-1}$ such that no simplex is completely labelled. Using Proposition 3.9 one can then create an $m \times n$ non-degenerate bimatrix game that does not possess an equilibrium, leading to a contradiction.

Take the division of $\triangle^{m-1}$ into labelled regions induced by the retraction r. Construct iteratively a triangulation of $\triangle^{m-1}$ such that its mesh is so small that no simplex is completely labelled. As before, the label of a vertex is a label of a region that contains the vertex. Note that the mesh of the triangulation can be constructed arbitrarily small (see Lemma 3.11)

Let $v_1, \ldots, v_N$ be the set of vertices added to the triangulation, where the subscript reflects the order in which the vertices are added. Let $\Lambda \subset \{1, \ldots, N\}$ denote the ordered subset for those vertices that were added to the boundary of $\triangle^{m-1}$. Now take the vertex v_λ, for $\lambda \in \Lambda$, that is added last to the triangulation, and consider the iterated refinement that is obtained by adding the set of vertices $\{v_1, \ldots, v_N\} - \{v_\lambda\}$ in canonical order. Continuing with the second-to-last vertex that was added to the boundary of $\triangle^{m-1}$ and so forth finally gives an iterated refinement with no vertices added to the boundary of $\triangle^{m-1}$ that, by Lemma 3.12, is regular (see also Lemma 4.2 in the next chapter).

It remains to show that the deletion of vertices on the boundary does not create completely labelled simplices. Let v be a vertex that was added to the boundary. Then $v = \sum_{i=1}^{l} \mu_i v_i$ with $\mu_i > 0$ and $1_l^\top \mu = 1$, for some $v_1, \ldots, v_l$. Note that the retraction r is the identity on the boundary of $\triangle^{m-1}$. In particular, the labelling satisfies $L(v) = L(v_i)$ for some $i \in \{1, \ldots, l\}$. So the face spanned by $\{v_1, \ldots, v_{i-1}, v, v_{i+1}, \ldots, v_k\}$ has the same labels as the face spanned by $\{v_1, \ldots, v_{i-1}, v_i, v_{i+1}, \ldots, v_k\}$. So a simplex spanned by $\{v_1, \ldots, v_{i-1}, v, v_{i+1}, \ldots, v_k\}$ and some $\{v_{k+1}, \ldots, v_m\}$ is fully labelled if and only if the simplex spanned by $\{v_1, \ldots, v_{i-1}, v_i, v_{i+1}, \ldots, v_k\}$ and $\{v_{k+1}, \ldots, v_m\}$ is fully labelled. Hence v can be removed without creating a completely labelled simplex (see also Lemma 4.4 in the next chapter).   □

McLennan and Tourky (2004) have recently shown how Kakutani's fixed point theorem can be proven by game theoretic concepts. They create games whose Nash equilibria yield approximate fixed points, where the existence of the Nash equilibria is ensured by the Lemke-Howson algorithm. The authors argue that "the Lemke-Howson algorithm embodies, in algebraic form, the fixed point principle itself, and not merely the existence theorem for finite two person games" (p. 3–4). The analysis above supports this view.

This section concludes with an observation that shows how to translate the labelled dual construction $X_*^\triangle$ into a labelled triangulation that satisfies the Sperner condition such that it reflects the combinatorial properties of $X_*^\triangle$.

Proposition 3.14. *Let $X_*^\triangle$ be the labelled dual construction for some $(m \times n)$-bimatrix game, and let $|X^\triangle|$ denote the regular triangulation of $X^\triangle$. Then there exists a labelled refinement of $|X^\triangle|$ such that a vertex in the refinement has label i if and only if it is contained in the region with label i and such that a simplex is completely labelled if and only if it contains a completely labelled point $w_s \in X_*^\triangle$.*

Proof. Take some simplex $v^\triangle$. The polyhedral division is generally not such that one can just label the vertices of $v^\triangle$ with the respective best reply labels without refining $v^\triangle$. Consider for example the polyhedral subdivisions depicted in Figure 3.10. In the first case, just labelling the vertices would yield a labelling such that the simplex is not completely labelled, although it contains a fully labelled point. In the second case, one would obtain a completely labelled simplex, although it does not contain a completely labelled point. Therefore, refinement is necessary.

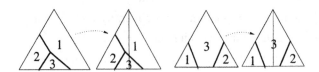

Fig. 3.10. A refinement of $v^\triangle$

Now one can refine the mesh of $|X^\triangle|$. This can, for example, be achieved by an iterated refinement. If the refinement is sufficiently small, a simplex contains a fully labelled point if and only if all its vertices lie in distinct best reply regions. Labelling the vertices according to the best reply region yields the desired labelled refinement.   □

A possible refinement for the game in Example 2.3 is depicted in Figure 3.11. In this case, it is sufficient to add a vertex to the edge between vertices representing strategies 4 and 7. The resulting refinement fulfils the requirements of Proposition 3.14.

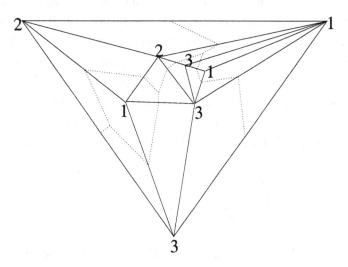

Fig. 3.11. A labelled triangulation for the game in Example 2.3

3.3 A Topological Interpretation of the Dual Construction

In the Sperner case above, a mapping f^S characterises the completely labelled simplices in the sense that a simplex is completely labelled if and only if it contains a point that is mapped to the completely labelled point $v_* \in \triangle_*^{m-1}$. This mapping can be described by the Sperner matrix $A^S(\triangle)$ for each simplex $\triangle$ in the triangulation. The aim of this section is to construct a similar mapping $f^\triangle$ for $X_*^\triangle$ via the artificial payoff matrix $A(v)$. This mapping is used in extending the dual construction to outside option equilibrium components and when giving a new characterisation of index $+1$ equilibria.

Take the payoff matrix A for player I. First the columns A_j of A, for $j \in J$, are normalised as follows. Without loss of generality it can be assumed that all entries of A_j are greater than zero. Otherwise one can add a positive constant to all payoffs without affecting the best reply regions and hence the equilibria of the game. Let $|A_j| = \sum_{i=1}^{N} A_{ij}$, i.e. $|A_j|$ denotes the sum of entries in column A_j. By assumption $|A_j| \neq 0$. Let $A_{max} = max_{j \in J} |A_j|$. Add the positive constant $\frac{A_{max} - |A_j|}{m}$ to column j. Adding a positive constant to a column of player I's payoff matrix also leaves the equilibria and best reply regions invariant. In the modified payoff matrix, the entries in each column add up to A_{max}. Now divide all payoffs by A_{max}. This, again, leaves the Nash equilibria invariant. Hence one obtains an equivalent payoff matrix, also denoted as A, in which all entries are positive and in which the column entries add up to $+1$.

Now consider a simplex $v^\triangle$ in $|X^\triangle|$. Let w_s be a point in $v^\triangle$. The point w_s can be described by convex coordinates with respect to the vertices of $v^\triangle$. So for a point w_s in $v^\triangle$ that is given by its coordinates with respect to the vertices $v^\triangle$ one can simply define $f_v(w_s) = A(v)w_s$. Then $f_v(w) \in \triangle^{m-1}$ since

$$|A(v)w| = \sum_i (A(v)w)_i = \sum_i \sum_j A(v)_{ij} w_j = \sum_j \sum_i A(v)_{ij} w_j$$
$$= \sum_j w_j \sum_i A(v)_{ij} = \sum_j w_j = 1.$$

A depiction of f_v is given in Figure 3.12. It shows a simplex $v^\triangle$ spanned by vertices v_1, v_2 and v_3 and its image in $\triangle_*^{m-1}$. The columns of $A(v)$ are given by A_1, A_2 and A_3. By construction, the columns A_i ($i = 1, 2, 3$) are elements of $\triangle^{m-1}$. So the image of f_v is the subset of $\triangle_*^{m-1}$ that is spanned by the payoff vectors A_1, A_2 and A_3 in $\triangle_*^{m-1}$. In particular, the image is some simplex that lies in $\triangle_*^{m-1}$ (this simplex is not necessarily full dimensional, even for non-degenerate payoff vectors). The division of $v^\triangle$ into best reply regions is an affine transformation of the division of the simplex spanned by A_1, A_2 and A_3, whose division is that induced by the division of $\triangle_*^{m-1}$.

If v_1 and v_2 share a common face, the mappings f_{v_1} and f_{v_2} are identical on that face. Hence, by defining f piecewise on each simplex $v^\triangle$ as f_v, one obtains a mapping

$$f: (X_*^\triangle, \partial X_*^\triangle) \longrightarrow (\triangle_*^{m-1}, \partial \triangle_*^{m-1}). \tag{3.3}$$

Note that the mapping on the boundary of $X^\triangle$ is given by the unit vectors as components of $A(v)$, so f maps boundary on boundary. Furthermore, by construction, the labels of a point w_s are the same as the labels of its image. The mapping f in (3.3) is referred to as the *payoff mapping*, since the value

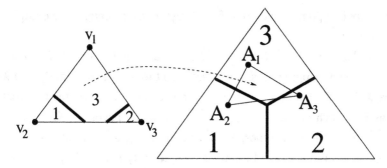

Fig. 3.12. The mapping f_v

of f is the expected payoff of player I under a strategy profile w_s of player II (including the slack variables). A depiction of the underlying geometry is given in Figure 3.13. It shows that the simplex marked in dashed lines is mapped affinely on a simplex in $\triangle_*^{m-1}$, also described by dashed lines. The vertices of the simplex in $\triangle_*^{m-1}$ are the images of the vertices in $|X^{\triangle}|$.

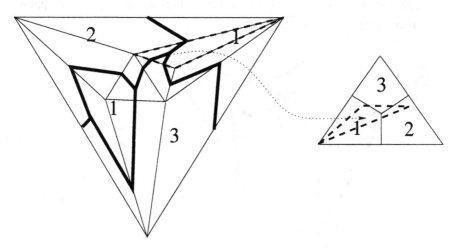

Fig. 3.13. The payoff mapping f

This is a crucial difference to the Sperner case. There, the images of simplices are either the simplex $\triangle_*^{m-1}$ itself (if the simplex is completely labelled), or the images are faces of $\triangle_*^{m-1}$ (if the simplex is not completely labelled). In the dual construction, the images of simplices $v^{\triangle}$ are simplices which are contained in $\triangle_*^{m-1}$. Nevertheless, the simplex $v^{\triangle}$ contains a completely labelled point if and only if its image under f contains the completely labelled point v_*.

Note that $X = \triangle^{m-1}$. So, so far, f is a mapping $f : X^\triangle \to X$. To define the index via a mapping, it is more convenient to have a mapping $X_*^\triangle \to X^\triangle$, where $X^\triangle$ is divided into best reply regions as in $P_*^\triangle$, i.e. via the unit matrix that assigns each vertex $-m\hat{v}e_i$ of $X^\triangle$ the artificial payoff e_i. The simplices $\triangle_*^{m-1}$ and $X^\triangle$ are homeomorphic via the mapping $\mathrm{Id}^\triangle$ that is described by the matrix $-m\hat{v} \cdot \mathrm{Id}$, where Id is the $m \times m$ identity matrix. In particular, the labels of a point $w \in \triangle_*^{m-1}$ are the same as the labels of its image $\mathrm{Id}^\triangle(w)$. This is due to the fact that the vertex in $\triangle^{m-1}$ with label i is mapped to the vertex of $X^\triangle$ with label i.

Using $\mathrm{Id}^\triangle$, one defines the *dual payoff mapping* $f^\triangle$ as the composition of $\mathrm{Id}^\triangle$ and f, i.e. $f = \mathrm{Id}^\triangle \circ f$. This yields

$$f^\triangle : (X_*^\triangle, \partial X_*^\triangle) \longrightarrow (X^\triangle, \partial X^\triangle) \tag{3.4}$$

A depiction of $f^\triangle$ is given in 3.14. The only difference to the payoff mapping f is that it maps $X_*^\triangle$ on $X^\triangle$ instead of $\triangle_*^{m-1}$.

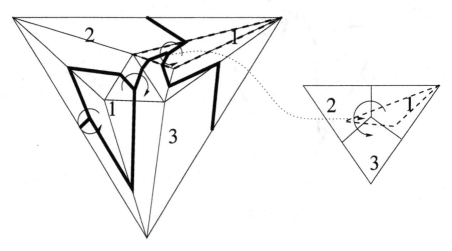

Fig. 3.14. The dual payoff mapping $f^\triangle$

The difference between $X^\triangle$ and $\triangle^{m-1}$ is that they have the same orientation relative to projection point $v_p = (-m\hat{v}, \ldots, -m\hat{v})$ for odd m, and opposite orientation for even m. This is depicted in Figure 3.15, and can be verified using an inductive argument.

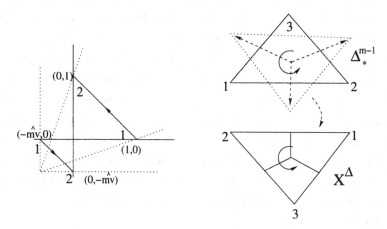

Fig. 3.15. The orientation of the $X^\triangle$ and $\triangle^{m-1}$

For notational convenience, let v_* denote the completely labelled point in $X^\triangle$ (as it does in $\triangle_*^{m-1}$). Note that both completely labelled points in $X^\triangle$ and $\triangle_*^{m-1}$ have coordinates $(\frac{1}{m}, \ldots, \frac{1}{m})$ with respect to the vertices of $X^\triangle$ and $\triangle_*^{m-1}$. So the equilibria of a game are represented by exactly those points w_s that are mapped to v_* under the mapping $f^\triangle$. Also, the index can be described by the local degree of $f^\triangle$.

Lemma 3.15. *Let $w_s \in (f^\triangle)^{-1}(v_*)$. Then the index of w_s as in Definition 2.9 is the same as the local degree of $f^\triangle$ at w_s.*

Proof. The index in Definition 2.9 is defined by a permutation of the labels I of a simplex $w_s^\triangle$, which corresponds to a permutation of vertices. For a mapping that permutes the vertices of a simplex, the degree equals the sign of the permutation (see e.g. Dold (1972, IV, 4, Example 4.3)). □

Using the mapping $f^\triangle$ and degree theory, it follows that the sum of indices over the equilibria of a game equals $+1$, so the number of equilibria is odd. This can be seen as follows. The degree of the mapping $f^\triangle$ has similar properties to the degree of the Sperner mapping f^S described on page 67. Similar to the Sperner mapping, the degree of the mapping $f^\triangle$ counts the number of completely labelled points in $X_*^\triangle$, where each point is counted with its local degree. This local degree is, by Lemma 3.15, the same as the index.

Furthermore, the degree of the mapping $f^\triangle$ is the same as the degree of $f^\triangle$ restricted to the boundary of $X_*^\triangle$. Similar to the Sperner mapping, the degree of $f^\triangle$ restricted to the boundary of $X_*^\triangle$ counts, for a fixed label $k \in I$, the number of almost completely labelled points on the boundary of $X_*^\triangle$ with labels $I - \{k\}$, counted by their local orientation. The orientation on the boundary is induced by the orientation of the boundary of $X^\triangle$. This number is independent of k. For each $k \in I$, there is exactly one point on the boundary of $X_*^\triangle$ with labels $I - \{k\}$. The local orientation of this point is $+1$ as it is contained in the face of $X^\triangle$ spanned by $-m\hat{v}e_i, i \in I - \{k\}$. Alternatively, one sees that $f^\triangle$ restricted to the boundary is the identity, and hence its degree is $+1$ (for a detailed account of degree theory see e.g. Dold (1972) as cited on p. 67).

4

A Strategic Characterisation of the Index

This chapter provides a new characterisation of the index for equilibria in non-degenerate bimatrix games in terms of a strategic property. It is shown that an equilibrium has index $+1$ if and only if one can add strategies with new payoffs to the game such that the equilibrium is the unique equilibrium of the extended game.

Suppose one can add strategies to a game such that an equilibrium remains the unique equilibrium of the extended game. Since the indices of equilibria of a game have to add up to $+1$, it follows that the equilibrium must have index $+1$ in the extended game. But the index only depends on the strategies played with positive probability, so it follows that the index of the equilibrium in the original game also equals $+1$. Hence, if one can extend the game such that the equilibrium becomes the unique equilibrium of the extended game, the index of that equilibrium must equal $+1$. Here it is shown that the converse is also true, i.e. if an equilibrium has index $+1$ then one can add strategies such that the equilibrium becomes the unique equilibrium of the extended game. This yields a new characterisation of the index purely in terms of a strategic property.

The structure of this chapter is as follows. Section 4.1 shows the result for the special case of pure strategy equilibria (Lemma 4.1) and motivates the general result by examining particular examples. Section 4.2 provides some technicalities that are also needed in Chapter 6. Section 4.3 shows that an equilibrium in a non-degenerate bimatrix game has index $+1$ if and only if one can add strategies to the game such that the equilibrium is the unique equilibrium of the extended game (Theorem 4.6). It turns out to be sufficient to just add strategies for one player.

4.1 A Geometric Interpretation

The properties of the index imply that the index of an equilibrium is $+1$ if one can add strategies such that the equilibrium becomes the unique equilibrium in the extended game. The indices of equilibria of a game have to add up to $+1$. So the index of a unique equilibrium in an extended game equals $+1$. But the index does only depend on strategies played with positive probability, and hence the index of the equilibrium in the original game equals $+1$.

Pure strategy equilibria in non-degenerate bimatrix games have index $+1$. For these it is easy to see that they can be made the unique equilibrium in some extended game.

Lemma 4.1. *Let G be an $m \times n$ non-degenerate bimatrix game. Then every pure strategy equilibrium of the game is the unique equilibrium in some extended game.*

Proof. Let G be represented by $m \times n$ payoff matrices A and B. Without loss of generality (otherwise one can reorder the strategies) assume that the pure strategy equilibrium is given by player I playing strategy 1 and player II playing strategy $m + 1$ (i.e. both play their first strategy). Then add strategy with label $m + n + 1$ for player II with payoff column, for small $\varepsilon > 0$,

$$
\begin{pmatrix}
1, b_{11} - \varepsilon \\
0, \max_{j=1,\dots,n} b_{2j} + \varepsilon \\
\vdots \\
0, \max_{j=1,\dots,n} b_{mj} + \varepsilon
\end{pmatrix}.
\tag{4.1}
$$

Then strategy $m + n + 1$ strictly dominates all other strategies except for strategy $m + 1$ of player II. Note that $b_{11} > b_{1j}$ for all $j \in J$, for $j \neq 1$. So strategies $j = m + 2, \dots, m + n$ can be deleted. Thereafter, strategy 1 strictly dominates all other strategies $2, \dots, m$ of player I. By iterated elimination of strictly dominated strategies, only the strategy pair $(1, m + 1)$ remains. $\qquad \square$

Adding strategies as in Lemma 4.1 alters the dual construction for the
game. Take, for example, game H^- as in (1.13). The game is given by

$$H^- = \begin{bmatrix} 13,13 & 7,12 & 1,14 \\ 12,7 & 8,8 & 2,1 \\ 14,1 & 1,2 & 1,1 \end{bmatrix}.$$

This game has three equilibria. The mixed equilibrium with index -1 in
which both players play $(\frac{1}{2},\frac{1}{2},0)$, the pure strategy equilibrium with index $+1$
in which both players play $(0,1,0)$, and the completely mixed equilibrium
with index $+1$ in which both players play $(\frac{1}{2},\frac{1}{12},\frac{5}{12})$. The labelled dual construction for the game is depicted on the left in Figure 4.1.

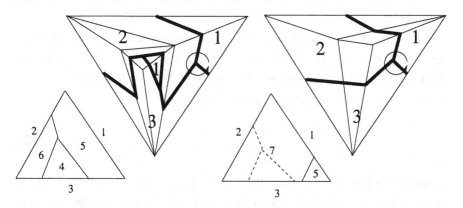

Fig. 4.1. An index $+1$ equilibrium in H^-

Now suppose the game is extended in the following way, so that only the
pure strategy equilibrium remains.

$$\tilde{H}^- = \begin{bmatrix} 13,13 & 7,12 & 1,14 & 0,20 \\ 12,7 & 8,8 & 2,1 & 10,7\frac{1}{2} \\ 14,1 & 1,2 & 1,1 & 0,20 \end{bmatrix}.$$

The added strategy dominates strategies 4 and 6 of player II. So strategies
4 and 6 can be deleted. Then strategy 2 of player I is the best reply to both
strategies 5 and 7, and the best reply to strategy 2 is 5. Thus the pure strategy
equilibrium in which player I plays strategy 2 and player II plays strategy 5

(with payoff 8 for both players) is the unique equilibrium of the extended game.

Adding strategies changes the dual construction for the game. Consider the labelled dual construction for the extension of the game (1.13), which is depicted on the right in Figure 4.1. The paths that start from the completely labelled point that represents the pure strategy equilibrium lead directly to the boundary. In the original game some paths in the dual construction lead to other equilibria of the game as shown on the left in Figure 4.1. So, in order to make an index $+1$ equilibrium the unique equilibrium of an extended game, the paths that start in the fully labelled point representing the equilibrium have to be "re-routed" such that they connect directly with the boundary of the dual construction, also not creating other equilibria (e.g. pairs of inaccessible equilibria).

The idea of "re-routing" the paths is the main idea in the proof of Theorem 4.6 below. To give the reader an idea of the process, the procedure is first applied to examples before it is technically specified in the proof of Theorem 4.6. Take for example the following game.

$$\begin{bmatrix} 1,3\ 0,2\ 1,0 \\ 0,0\ 1,2\ 0,3 \end{bmatrix}. \tag{4.2}$$

Game (4.2) has 3 equilibria. The pure strategy equilibrium $(1,0),(1,0,0)$ with index $+1$, the mixed equilibrium $(\frac{2}{3},\frac{1}{3}),(\frac{1}{2},\frac{1}{2},0)$ with index -1, and the mixed equilibrium $(\frac{1}{3},\frac{2}{3}),(0,\frac{1}{2},\frac{1}{2})$ with index $+1$. The dual construction for this game is given on the left in Figure 4.2 (the dots represent the vertices of the simplices $v^\triangle$).

Now suppose one wants to make the equilibrium $(\frac{1}{3},\frac{2}{3}),(0,\frac{1}{2},\frac{1}{2})$ the unique equilibrium of an extended game. The dual construction shows how to achieve this. Add a strategy 6 for player II, covering the best reply region of strategy 3 and a small part of the best reply region of strategy 4. This can, for example, be achieved by choosing the payoff vector $\binom{4}{0}$ for player II. The new division of X and its dual are depicted on the right in Figure 4.2. Then choose strategy 2 to be the best reply to the new strategy 6 by, for example, choosing the payoff vector $\binom{0}{1}$ for player I. Then $(\frac{1}{3},\frac{2}{3}),(0,\frac{1}{2},\frac{1}{2},0)$ is the unique equilibrium of the extended game

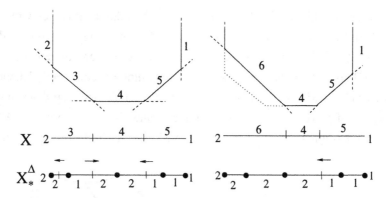

Fig. 4.2. An index $+1$ equilibrium for $m = 2$

$$\begin{bmatrix} 1,3 & 0,2 & 1,0 & 0,4 \\ 0,0 & 1,2 & 0,3 & 1,0 \end{bmatrix}. \tag{4.3}$$

The orientation around an index $+1$ equilibrium in the labelled dual construction agrees with the orientation of $X^\triangle$. This allows one to "re-label" the regions in the dual construction by adding strategies such that the index $+1$ equilibrium remains the unique equilibrium in the extended game. For any $2 \times n$ game the procedure is very straightforward and easy. It can easily be verified that one only has to add at most two strategies for player II to make any index $+1$ equilibrium the unique equilibrium in an extended game.

In higher dimensions, the process of eliminating the other equilibria without creating new equilibria is more advanced. Consider, for example, the following 3×3 coordination game.

$$\begin{bmatrix} 10,10 & 0,0 & 0,0 \\ 0,0 & 10,10 & 0,0 \\ 0,0 & 0,0 & 10,10 \end{bmatrix}. \tag{4.4}$$

Game (4.4) is the same as the game H^3 given by (1.16). All three pure strategy equilibria have index $+1$, the three mixed equilibria with two strategies as support have index -1, and the completely mixed equilibrium has index $+1$ again. Making a pure strategy equilibrium of (4.4) the unique equilibrium in an extended game is straightforward (see Lemma 4.1). So suppose one wants to make the completely mixed equilibrium the unique equilibrium of some extended game. In order to do so, one first has to cover the old equilibria with

new strategies. This can be done, for example, by adding strategies with labels 7, 8 and 9 for player II as shown in Figure 4.3. In a neighbourhood of the
vertex $v = (\frac{1}{3}, \frac{1}{3}, \frac{1}{3}) \in X$, the structure of the best reply regions remains unchanged. This implies that the simplex $v^\triangle$ containing the completely labelled
point remains unaffected by the added strategies. This first step determines
the payoffs of player II for the added strategies and gives a triangulation $|X^\triangle|$
in which the original simplex $v^\triangle$ and its division are as in the original game.

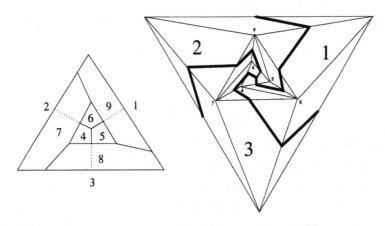

Fig. 4.3. A unique index $+1$ equilibrium in an extension of the coordination game

Second, one has to choose the appropriate payoffs for player I. The right of
Figure 4.3 shows how the paths starting in the corresponding dual of the equilibrium can be "re-routed". So the payoffs for player I are chosen in such a
way that the almost completely labelled points on the boundary of $v^\triangle$ are connected with the respective almost completely labelled points on the boundary
of the dual. The game that corresponds with the labelled dual on the right in
Figure 4.3 is given by

$$\begin{bmatrix} 10,10 & 0,0 & 0,0 & 0,11 & 10,5 & 0,-10 \\ 0,0 & 10,10 & 0,0 & 0,-10 & 0,11 & 10,5 \\ 0,0 & 0,0 & 10,10 & 10,5 & 0,-10 & 0,11 \end{bmatrix}. \qquad (4.5)$$

So, in order to prove that an index $+1$ is the unique equilibrium in some
extended game, one essentially has to show two things. First, that the paths
can in fact be re-routed. This is ensured by the index $+1$ condition. Second,

one has to show that these paths can actually be created by extending the game. This is to say that in the labelled dual construction of the extended game the paths starting in the equilibrium connect directly with the boundary. Adding columns to the payoff matrix B refines the mesh of $|X^\triangle|$, and the payoffs for player I determine the paths.

4.2 Some Technical Requisites

The proof of Theorem 4.6 below is based on the approximation of a homotopy that "re-routes" the paths. In order to show that the approximation of the homotopy can be achieved by adding strategies, this section provides some technical results that are required in the proof of Theorem 4.6. These technical results are also used in the characterisation of index zero outside option equilibrium components in Chapter 6.

Let $\triangle$ be an $(m-1)$-simplex in a regular triangulation $|\triangle^{m-1}|$ of $\triangle^{m-1}$ with no vertices on the boundary of $\triangle^{m-1}$ other than e_i, $i \in I$. Now consider an iterated refinement of $|\triangle^{m-1}| - \triangle$ that is achieved by subsequently adding vertices to $|\triangle^{m-1}| - \triangle$, allowing to add vertices on the boundary of $|\triangle^{m-1}|$ or $\triangle$. Let the added vertices be denoted as $v_1, \ldots, v_N$, where the subscript denotes the order in which the vertices were added. Now add the simplex $\triangle$. The resulting object is a division of $|\triangle^{m-1}|$ into simplices that is not a triangulation of $|\triangle^{m-1}|$. Such a division of $|\triangle^{m-1}|$ is referred to as an *iterated pseudo refinement*. An illustration of an iterated pseudo refinement is given in Figure 4.4.

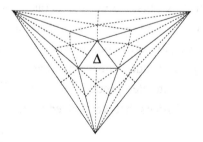

Fig. 4.4. An iterated pseudo refinement

Lemma 4.2. *Given an iterated pseudo refinement of $\triangle^{m-1}$, one can subsequently delete those vertices that were added to the boundary of $\triangle$ and $\triangle^{m-1}$ in order to obtain a regular refinement of $|\triangle^{m-1}|$.*

Proof. Let $v_1,\ldots,v_N$ be the set of vertices added to the triangulation, where the subscript reflects the order in which the vertices are added. Let $\Lambda \subset \{1,\ldots,N\}$ denote the ordered subset for those vertices that were added to the boundary of $\triangle$ or $\triangle^{m-1}$. Now take the vertex v_λ, for $\lambda \in \Lambda$, that is added last to the triangulation, and consider the iterated pseudo refinement that is obtained by adding the set of vertices $\{v_1,\ldots,v_N\} - \{v_\lambda\}$ in canonical order. Continuing with the second last vertex that was added to the boundary of $\triangle$ or $\triangle^{m-1}$ and so forth, finally gives an iterated pseudo refinement with no vertices added to the boundary of $\triangle$ or $\triangle^{m-1}$. Hence, the refinement achieved by adding the set of vertices $\{v_1,\ldots,v_N\} - \{v_\lambda \mid \lambda \in \Lambda\}$ (in canonical order) is regular by Lemma 3.12. $\qquad\square$

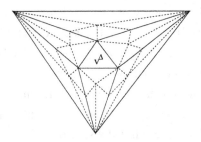

Fig. 4.5. The regular refinement obtained from the iterated pseudo refinement

The refinement that is obtained by the iterated pseudo refinement in Figure 4.4 is depicted in Figure 4.5. The result of Lemma 4.2 extends in a straightforward way to collections of simplices $\bigcup_i \triangle_i$ in a triangulation $|\triangle^{m-1}|$ and iterated pseudo refinements that are obtained by refining $|\triangle^{m-1}| - \bigcup_i \triangle_i$. So every iterated pseudo refinement yields a regular refinement by omitting those vertices that were added to the boundary of $\bigcup_i \triangle_i$ or $\triangle^{m-1}$.

Now consider an iterated pseudo refinement of $|X^\triangle| - v^\triangle$. Vertices that were added to the boundary of $X^\triangle$ or $v^\triangle$ are referred to as *pseudo vertices*. Assign a payoff vector A_{v_i} to each added vertex v_i. If the added vertex is a pseudo vertex, then the payoff vector is referred to as a *pseudo payoff vector*. Each pseudo vertex $\tilde{v}$ can be described as a convex combination of $m-1$ vertices $v_1, \ldots, v_{m-1}$ on the boundary of $X^\triangle$ or the boundary of $v^\triangle$, i.e. $\tilde{v} = \sum_{i=1}^{m-1} \mu_i v_i$, with $\mathbf{1}_m^\top \mu = 1$ and $\mu_i \geq 0$.

Definition 4.3. *The pseudo payoffs are called* consistent *if $A_{\tilde{v}} = \sum_{i=1}^{m-1} \mu_i A_{v_i}$.*

For each simplex in the pseudo refinement of $|X^\triangle| - v^\triangle$, the payoff vectors and pseudo payoff vectors induce a division into labelled regions as described by (2.7), where the columns of the payoff matrix consist of the payoff vectors and pseudo payoff vectors that are assigned to the vertices of the simplex. This division is referred to as a *pseudo division*.

Now consider the regular refinement induced by an iterated pseudo refinement. The following lemma is similar to what was used in the proof of Corollary 3.13. That is, if the pseudo vectors have consistent payoffs, then the induced division of $|X^\triangle| - v^\triangle$ into labelled regions is unaffected by deleting the pseudo vectors from the iterated pseudo refinement.

Lemma 4.4. *If the pseudo payoffs are consistent, then the pseudo division of $|X^\triangle| - v^\triangle$ into labelled regions is identical with the division of $|X^\triangle| - v^\triangle$ into labelled regions that is obtained by deleting the pseudo vertices from the iterated pseudo refinement.*

Proof. The proof is illustrated in Figure 4.6. The consistency of the payoff ensures that the division of a larger simplex is given by the division of the smaller simplices. In the figure, the payoff for v is consistent with the payoffs for v_1 and v_2. Then the union of the simplices spanned by $\{v_1, v_2, v\}$ and $\{v_2, v_3, v\}$ yields the same division as the simplex spanned by $\{v_1, v_2, v_3\}$.

Let v denote the simplex that was last added to the face of $v^\triangle$ or $X^\triangle$. Then $v = \sum_{i=1}^{k} \mu_i v_i$, with $\mathbf{1}_m^\top \mu = 1$ and $\mu_i > 0$, where the vertices v_i span the $(k-1)$-simplex on the $(m-2)$-face that contains v. These vertices might be original vertices or pseudo vertices. In any case, one has $A_v = \sum_{i=1}^{k} \mu_i A_{v_i}$. Now delete v from the iterated pseudo refinement. Consider a simplex $\triangle$ spanned by $v_1, \ldots, v_k$ and some $v_{k+1}, \ldots v_m$. The division of $\triangle$ is induced by the payoff vectors $A_{v_1}, \ldots, A_{v_m}$.

The simplex $\triangle$ is the union of smaller simplices for which the vertex v replaces one of the vertices v_i, $1 \leq i \leq k$, of $\triangle$. Since the payoffs are consistent,

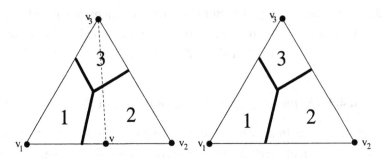

Fig. 4.6. Pseudo vertices with consistent payoffs

the induced division of $\triangle$ into labelled regions is also the same as the union of the smaller simplices divided into labelled regions. □

Finally, one needs a topological lemma, which says that the payoff mapping f (as in (3.3)) restricted to the boundary of $v^\triangle$ can be deformed into a mapping that maps the boundary of $v^\triangle$ on the boundary of $\triangle_*^{m-1}$.

Lemma 4.5. *Let $v^\triangle$ be a simplex in $|X^\triangle|$. Then there exists a homotopy h that deforms f (or $f^\triangle$) restricted to the boundary of $v^\triangle$ into a mapping that maps the boundary of $v^\triangle$ on the boundary of $\triangle_*^{m-1}$ (or the boundary of $X^\triangle$). The homotopy is such that $h(x,t) \neq v_* \ \forall (x,t) \in \partial v^\triangle \times [0,1]$.*

Proof. Take a simplex $v^\triangle$ in $|X^\triangle|$, and let $\partial v^\triangle$ denote its boundary. If the image of $v^\triangle$ contains v_*, then v_* must lie in the interior of $f(v^\triangle)$. If the image does not contain v_*, then v_* must have a positive distance from $f(v^\triangle)$. This is due to the non-degeneracy assumption.

Then one can retract the image of the boundary $f(\partial v^\triangle)$ as follows: Let x be a point on $f(\partial v^\triangle)$. Then take the line between x and v_* in direction of x, and define the retraction $r(x)$ as the point on the boundary of $\triangle_*^{m-1}$ in which the line intersects with the boundary of $\triangle_*^{m-1}$. Algebraically, the point $r(x)$ is the normalised form of the vector $x - (min_{i \in I} x_i) \cdot \mathbf{1}_m$. The retraction $r(x)$ can be described as a homotopy $h : \partial v^\triangle \times [0,1] \to \triangle_*^{m-1}$ given by $h(x,t) = t \cdot r(x) + (1-t) \cdot x$. Note that $h(x,t) \neq v_* \ \forall \ (x,t) \in \partial v^\triangle \times [0,1]$, since x and $r(x)$ have the same labels.

A deformation of f restricted to $\partial v^\triangle$ yields a deformation of $f^\triangle$ restricted to $\partial v^\triangle$, since $f^\triangle = \mathrm{Id}^\triangle \circ f$. □

Lemma 4.2 and 4.4 are needed in the proof of Theorem 4.6 below. In the proof, a certain mapping is approximated. For this one needs to construct a triangulation with a sufficiently small mesh. This can only be achieved by adding vertices to certain boundary faces. However, if the payoffs are consistent, then these vertices can be omitted, as it does not change the combinatorial division into best reply regions. In particular, one obtains a regular triangulation and a division into labelled regions that can be obtained as the dual construction for some bimatrix game. Lemma 4.5 is needed to construct the mapping that is approximated.

4.3 A Game Theoretic Characterisation of the Index

This section proves the main result of this chapter, i.e. an equilibrium in a game has index $+1$ if and only if one can add strategies to the game such that the equilibrium becomes the unique equilibrium in the extended game. The idea of the proof is to "re-route" the paths as described earlier. Say (v, w) is an equilibrium. In the labelled dual construction, this equilibrium is represented by some $w_s \in v^\triangle$. In particular, if the index of the equilibrium is $+1$, the dual payoff mapping $f^\triangle$ restricted to the boundary of $v^\triangle$ has also degree $+1$. By a well-known result from algebraic topology, $f^\triangle$ restricted to the boundary of $v^\triangle$ and $f^\triangle$ restricted to the boundary of the $X_*^\triangle$ are homotopic via some homotopy h. This allows one to "re-route" the paths starting in w_s so as to connect them directly with the boundary without creating new equilibria.

Theorem 4.6. *Let G be some non-degenerate bimatrix game. Let $(v, w) \in X \times Y$ be an equilibrium of the game. Then (v, w) has index $+1$ if and only if one can add finitely many strategies such that (v, w) is the unique equilibrium of the extended game. It suffices to add strategies for only one player.*

Proof. Let $(v, w) \in X \times Y$ be an equilibrium of the game. First, all unplayed strategies of player II can be eliminated by new strategies that dominate them. If pure strategy $j \in J$ is not played in equilibrium, one can add a pure strategy j' with payoff $B_j + \varepsilon$, where $\varepsilon \in \mathbb{R}^m$ is a vector with small positive entries. This replaces the original vertex in $|X^\triangle|$ representing strategy j with a vertex representing the new strategy j'. In the dual polytope $P^\triangle$, this corresponds to adding a vertex to the boundary of $P^\triangle$ that lies slightly above the original vertex. This yields the same regular triangulation $|X^\triangle|$ as before.

Now consider the boundary of $v^\triangle$. Without loss of generality assume that all payoffs for player I are positive and that the payoffs in the columns of A add up to 1, i.e. $|A_j| = 1$ for $j \in J$ as assumed in the construction of $f^\triangle$. Let (v, w) be an equilibrium and consider the restriction of $f^\triangle$ to $v^\triangle$. Denote this restriction as $f^\triangle_{|v^\triangle}$.

The degree of the equilibrium is given by the local degree of $f^\triangle_{|v^\triangle}$ around the completely labelled point w_s, where w_s denotes the lifted point of w. The local degree is the same as the degree of $f^\triangle_{|v^\triangle}$ restricted to the boundary of $v^\triangle$, denoted as $f^\triangle_{|\partial v^\triangle}$, and has degree $+1$. The degree of $f^\triangle$ restricted to the boundary of $X^\triangle$, denoted as $f^\triangle_{|\partial X^\triangle}$, is also $+1$. Considering the payoff mapping f instead of the dual payoff mapping, this implies that $f_{|\partial v^\triangle}$ and $f_{|\partial X^\triangle}$ are homotopic (see e.g. Spanier (1966, 7.5.7)). First retract $f_{|\partial v^\triangle}$ to the boundary of $\triangle^{m-1}_*$ as shown in Lemma 4.5, then deform it into $f_{|\partial X^\triangle}$ along $\partial\triangle^{m-1}_*$. The construction is such that no point along the homotopy is mapped on v_*.

Denote this homotopy as h. The homotopy h is given as $h : \partial\triangle^{m-1}_* \times [0,1] \to \triangle^{m-1}_*$ such that $h(\cdot, 0) = f_{|\partial v^\triangle}$ and $h(\cdot, 1) = f_{|\partial X^\triangle}$. If $v^\triangle$ shares a common k-face with $X^\triangle$ (i.e. not all strategies of player I are played with positive probability in v), then the mappings $f_{|\partial v^\triangle}$ and $f_{|\partial X^\triangle}$ agree on that face by construction, and it can be assumed that $h(x, \cdot) = f_{|\partial v^\triangle}(x)$ for points x on that face.

But this gives a mapping, also denoted as h, on the space $X^\triangle - v^\triangle$ that agrees with f on the boundaries of $X^\triangle$ and $v^\triangle$ and whose image does not contain v_*. So

$$h : X^\triangle - v^\triangle \longrightarrow \triangle^{m-1}_*. \qquad (4.6)$$

This yields a division of $X^\triangle - v^\triangle$ into labelled regions such that no point is completely labelled. The regions are defined as the pre-images of the regions in $\triangle^{m-1}_*$. The division of $v^\triangle$ is as before. This is depicted in Figure 4.7 for the equilibrium (v_1, w_1) in the game of Example 2.3.

Now consider the triangulation $|X^\triangle|$, and consider an iterated pseudo refinement of $|X^\triangle| - v^\triangle$. This iterated pseudo refinement can be assumed to be such that no simplex has a diameter more than some $\delta > 0$ (see Lemma 3.11). Now assign payoffs for player I to the added vertices according to $A_v = h(v)$.

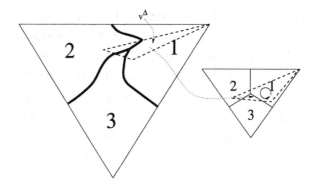

Fig. 4.7. A homotopy

If the simplices are small, their images in $\triangle_*^{m-1}$ are also small simplices (h is uniformly continuous), and no simplex contains v_*. This is depicted in Figure 4.8.

The pseudo payoffs for vertices that were added to the boundaries of $X^\triangle$ and $v^\triangle$ are consistent with the payoffs for the vertices of $X^\triangle$ and $v^\triangle$. Therefore, these vertices can safely be omitted without creating fully labelled points according to Lemma 4.4, and the resulting refinement is regular by Lemma 4.2. This refinement is a regular triangulation and can be achieved by a payoff matrix where strategies for player II are added (Lemma 3.12). The refinement determines the payoffs for player II. The payoffs for player I are given by the homotopy h. $\square$

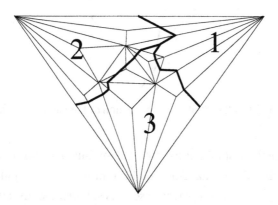

Fig. 4.8. An approximation of the homotopy

In the proof of Theorem 4.6, the simplices in the refinement are chosen to be sufficiently small since the homotopy h is not further specified. It is likely that, in the case of the payoff mapping f, one can easily describe the deformation of f restricted to the boundary, especially if considering the combinatorial aspects of the problem (instead of describing it as a topological problem). Furthermore, one is not necessarily bound to refining $|X^\triangle|$, but can actually create a new regular triangulation that leaves the simplex $v^\triangle$ unaffected. So, instead of adding sufficiently many strategies, it is likely that "a few" added strategies are enough.

As for the equilibrium (v_1, w_1) of the game in Example 2.3, it is sufficient to just add one strategy instead of many as suggested by Figure 4.8. The game described below only has the equilibrium (v_1, w_1) as a unique equilibrium.

$$\begin{bmatrix} 0,0 & 10,10 & 0,0 & 10,-10 & 0,11 \\ 10,0 & 0,0 & 0,10 & 0,8 & 1,1 \\ 8,10 & 0,0 & 10,0 & 8,8 & 0,1 \end{bmatrix}$$

Figure 4.9 depicts the corresponding labelled dual for the extended game.

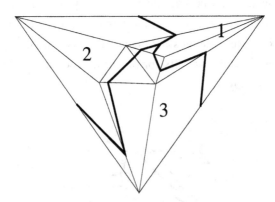

Fig. 4.9. The labelled dual for an extension of the game in Example 2.3

So the natural question arises about the minimal number of strategies one needs to add in order to make an equilibrium the unique equilibrium of an extended game. In the $2 \times n$ player case, it is sufficient to just add two strategies for player II to make any index $+1$ equilibrium the unique equilibrium of an extended game. Whether adding m or $2m$ strategies suffices in higher dimensions is unclear.

Remark 4.7. Instead of considering the homotopy h on $X^\triangle - v^\triangle$, one can actually define it on the "cylinder" that is obtained by deleting $X^\triangle$ and $v^\triangle$ from the surface of the polar polytope $P^\triangle$ that corresponds to the game.

Hofbauer (2000) defines two pairs $(G, (v, w))$, $(G', (v', w'))$, where (v, w) is an equilibrium of G, and (v', w') is an equilibrium of G', equivalent if the game G restricted to the support of (v, w) is the same as the game G' restricted to the support of (v', w'). He calls an equilibrium (v, w) of a game G sustainable if there exists an equivalent pair $(G', (v', w'))$ such that (v', w') is the unique equilibrium of G'. He conjectures that an equilibrium has index $+1$ if and only if it is sustainable. The results from above prove this conjecture in the case of non-degenerate bimatrix game.

5

Outside Option Equilibrium Components

The aim of this chapter is to extend the dual construction to outside option equilibrium components. This yields a new interpretation of the index for outside option equilibrium components that is very similar to a generalisation of Sperner's Lemma which is in the literature referred to as the *Index Lemma* (see e.g. Henle (1994), p. 47). The Index Lemma applies to more general boundary conditions, and states that the sum of orientations of completely labelled simplices can be deduced from the boundary condition. This new approach allows a new characterisation of index zero outside option equilibrium components in bimatrix games, which is the subject of Chapter 6.

An outside option can be thought of as an initial move that a player can make which terminates further play, and gives a constant payoff to both players. If the player has not chosen his outside option, the original game is played. Take for example the game described in (1.15) in Chapter 1. A representation of the game G^2 is given in Figure 5.1, where the bottom left entries in a cell are the payoff for player I and the top right entries in a cell are the payoffs for player II. This game has two equilibrium components: The single equilibrium of H^- with payoff 10 to both players, and the outside option equilibrium component with payoff 9 for player II and payoff 0 for player I.

In terms of *forward induction* the only reasonable equilibrium is that with payoff 10. Not playing *Out* in the first place is only reasonable if player II plays the equilibrium strategy that yields payoff 10 in H^-. Player I knows this and plays accordingly once the game H^- is entered. The notion of forward induction is a concept that applies to extensive form games (van Damme

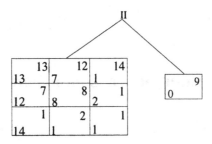

Fig. 5.1. A representation of an outside option game

(1989)). Other authors, in particular Kohlberg and Mertens (1986), argue that games should be analysed in their normal form and that solution concepts should be independent of the representation of the game. The index of an equilibrium component is an invariant, i.e. the same in all equivalent games and hence independent of the representation of the game. Therefore, understanding the nature of the index for outside option equilibrium components can help in understanding which solution concepts might capture the notion of forward induction (see e.g. Hauk and Hurkens (2002)). In Chapter 6, it is shown that an outside option equilibrium component is hyperessential if and only if it has non-zero index. It follows that an outside option outcome cannot be hyperessential if the forward induction equilibrium is a pure strategy equilibrium that is strict (that is, all unplayed pure strategies have a payoff that is strictly lower than the equilibrium payoff).

The structure of this chapter is as follows. Section 5.1 reviews a generalisation of Sperner's Lemma which is sometimes referred to as the Index Lemma (Proposition 5.2). In Section 5.2 it is shown how this relates to outside option equilibrium components (Corollary 5.4). Section 5.3 discusses potential generalisations and the apparent limitations of the dualisation method regarding general components of equilibria.

5.1 A Generalised Version of Sperner's Lemma

In Sperner's Lemma, the existence of a completely labelled simplex is ensured by the Sperner condition. Moreover, accounting for the orientation, the boundary condition determines that there exists one more completely labelled simplex with orientation $+1$ than with orientation -1. In this section, it is shown how Sperner's Lemma can be extended to cope with more general boundary conditions. This yields a generalisation of Sperner's Lemma that is in the literature referred to as the Index Lemma (see e.g. Henle (1994, p. 47)).

Let P be an $(m-1)$-dimensional polytope. Furthermore, let $|P|$ be a triangulation of P into simplices of dimension $m-1$. A triangulation of P is a finite collection of simplices whose union is P, and that is such that any two of the simplices intersect in a face common to both, or the intersection is empty. A triangulation of P induces a triangulation $|\partial P|$ of the boundary ∂P into simplices of dimension $m-2$. Let L be a labelling of the vertices of $|P|$ with labels in $I = \{1,\dots,m\}$. As before, one can define a Sperner mapping

$$f^S : (|P|,|\partial P|) \longrightarrow (\triangle_*^{m-1}, \partial\triangle_*^{m-1}),$$

where $\triangle_*^{m-1}$ denotes the canonical division described in Chapter 3 (see Definition 3.4): Every vertex of $|P|$ is mapped to the vertex in $\triangle_*^{m-1}$ with the corresponding label, i.e. $L(v) = L(f^S(v))$. Then f^S is obtained by linearly extending it to the simplices in $|P|$. Note that if a $(k-1)$-simplex has $j \le k$ distinct labels $I_j \subset I$, then it is mapped on the $(j-1)$-face of $\triangle_*^{m-1}$ that is spanned by the vertices with labels I_j. The restriction of f^S to the boundary of P is denoted as $f_{|\partial P}^S$.

Definition 5.1. *The index of the labelling L of $|P|$ is defined as*

$$I(L) = \deg f_{|\partial P}^S , \qquad (5.1)$$

where $\deg f_{|\partial P}^S$ *denotes the degree of the mapping* $f_{|\partial P}^S$.

As for the Sperner case, the degree $\deg f_{|\partial P}^S$ measures, for an arbitrary but fixed label $k \in I$, the number of almost completely labelled points with labels $I - \{k\}$ on the boundary, where each such point is counted with its orientation. The orientation on the boundary is induced by $\triangle_*^{m-1}$. This is depicted in Figure 5.2. The dotted line represents the image of the boundary ∂P "around" $\partial\triangle_*^{m-1}$. The mapping in Figure 5.2 has degree $+1$. The image of the boundary is homotopic to a single winding around $\triangle_*^{m-1}$. So the index of the labelling in Figure 5.2 is $+1$.

The degree $\deg f_{|\partial P}^S$ on the boundary is the same as the degree $\deg f^S$ of the mapping f^S. The proof of this claim is equivalent to the construction in the

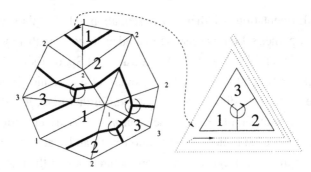

Fig. 5.2. A general version of Sperner's Lemma

proof of Theorem 3.3. There, the orientations of $(m-2)$-faces in the interior cancel out. The degree f^S measures the number of completely labelled points, i.e. the pre-images of v_*, where each pre-image is counted with its orientation, which is the local degree (see Figure 5.2). This fact that $\deg f^S_{|\partial P}$ is the same as $\deg f^S$ yields the following, well-known result, which says that the labelling of the vertices on the boundary determines the number of completely labelled simplices in the triangulation (for a detailed account of degree theory see e.g. Dold (1972) as cited on p. 67).

Proposition 5.2 (Index Lemma). *Let $|P|$ be as above with labelling L. Then the sum of orientations of the completely labelled simplices in $|P|$ equals $I(L)$.*

Proof. The pre-images of v_* correspond to the completely labelled simplices, and the local degree at a pre-image is the same as the orientation of the simplex that contains it. The degree equals the sum of local degrees, and is determined by the boundary condition.

Alternatively, one can use the same approach as in the proof of Theorem 3.3 to obtain the result without using degree theory. In this case, one would essentially show that $\deg f^S_{|\partial P}$ on the boundary is the same as the degree $\deg f^S$. $\qquad\square$

The Index Lemma is sometimes summarised with the phrase "The index equals the content" (see e.g. Henle (1994, p. 47)), meaning that the boundary condition (i.e. the index) determines the number of completely labelled simplices in the triangulation (i.e. the content), accounting for orientation. In the next section, it is shown that a similar description applies to outside option equilibrium components.

5.2 The Index for Outside Option Equilibrium Components

In Chapter 3 above it is shown how the classical Sperner condition applies to equilibria in non-degenerate bimatrix games. This section demonstrates how the Index Lemma relates to components of equilibria. The dual construction shows that the index of a component is defined by a boundary property similar to the Index Lemma. This boundary property determines the sum of indices of equilibria close to the component if the game is generically perturbed by small generic perturbations. In particular, it is shown that the sum of indices of equilibria close to the component is independent of the perturbation. This "invariance" property of the index for components of equilibria is not a new result (see the properties for components of equilibria listed in Section 1.3). What is new, however, is the geometric-combinatorial view on the index for components of equilibria.

The analysis is restricted to generic outside option equilibrium components in bimatrix games represented in strategic form by payoff matrices A and B. Without loss of generality it is assumed that the player with the outside option is player II. When player II plays the outside option, the payoffs for player I and player II are independent of player I's strategy choice. So the column of A that represents the payoffs for player I in the outside option has identical entries, and so has the column of B that represents the payoffs for player II in the outside option. An outside option equilibrium component is referred to as *generic* if the payoffs for player II are generic and if all payoffs for player I other than the outside option payoffs are generic. Thus the only degeneracy of the game arises through the payoffs to player I in the outside option. This implies that the payoffs for the equilibria that are cut off by the outside option are strictly smaller than the payoff in the outside option.

When constructing components of equilibria via outside options (see Section 1.4), it is possible to compute the index of such components purely on grounds of basic properties of the index. In particular, one does not have to go into details regarding the geometric-combinatorial aspects. These aspects, nevertheless, play an important role in the characterisation of index and (hyper)essentiality in Chapter 6. The examples given below are meant to illustrate the geometry behind the index for outside option equilibrium components by means of the labelled dual construction $X_*^\triangle$. A formal definition is given later in this section.

The problem with degenerate games is that, instead of having singleton solutions, one has to consider components of equilibria. This is due to the fact that the number of best reply strategies is not bounded by the size of the support (see Definition 1.1). In the case of an outside option in an $m \times n$ bimatrix game with an outside option for player II, the pure strategy representing the outside option for player II has m pure best reply strategies since all the payoffs for player I are the same in the outside option. In this case, the outside option equilibrium component C is given by

$$C = \{(x, Out) \in X \times Y \mid Out \text{ is best reply to } x\},$$

where Out denotes the pure strategy that represents the outside option.

In general, the dual construction cannot be applied to degenerate games. This is due to the fact that $|X^\triangle|$ is not well-defined if the payoff matrix B is degenerate. In the case of generic outside options in bimatrix games, however, the payoff matrix B is generic, since it does not matter if a column of B has identical entries. This allows one to apply the dual construction to such games. Consider, for example, the following 3×4 coordination game with an outside option for player II:

$$\begin{bmatrix} 10,10 & 0,0 & 0,0 & 0,9 \\ 0,0 & 10,10 & 0,0 & 0,9 \\ 0,0 & 0,0 & 10,10 & 0,9 \end{bmatrix}. \tag{5.2}$$

This is the same game G^{-2} in (1.17) in Chapter 1. The outside option equilibrium component has index -2. The three pure strategy equilibria of the game with payoff 10 (which are not cut off by the outside option) each have index $+1$. Since the sum of indices over all equilibrium components must equal $+1$, the outside option equilibrium component has index -2. This can be interpreted geometrically in the following way. Label the strategies of player I with $1, 2$ and 3, and those of player II with $4, 5, 6$ and Out. Then apply the dual construction to X to obtain $X_*^\triangle$. Figure 5.3 shows the division of X into best reply regions and the corresponding labelled dual construction $X_*^\triangle$ on the top. Strategy *Out* yields a constant payoff to player I. Therefore, the best reply regions in simplices $v^\triangle$ for which a vertex of $v^\triangle$ represents *Out* all join in the vertex that represents *Out*.

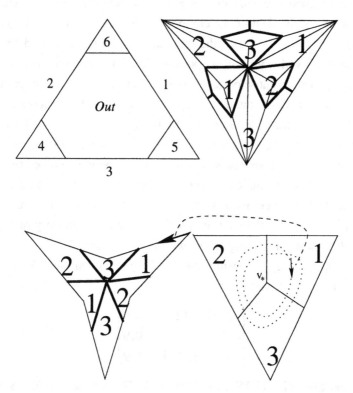

Fig. 5.3. An outside option component with index -2

The dual payoff mapping $f^\triangle$ as in (3.4) is, however, well-defined on $X^\triangle$, including those simplices that are the duals of the vertices of the best reply region for *Out*. In particular, the dual payoff mapping $f^\triangle$ is well-defined on the boundary of the dual of the outside option component.

The *dual of the outside option component* is the union of all those simplices that are the duals of the vertices of the best reply region for *Out*. These are the simplices that have *Out* as a vertex. The vertex that represents *Out* has all labels, since every strategy of player I is a best reply against *Out*. In particular, the completely labelled point does not lie in the interior of a simplex, which would be the case for non-degenerate bimatrix games. This is depicted on the bottom in Figure 5.3.

The dual of the component can now be used to define the index of an equilibrium component. For this, consider the dual payoff mapping restricted to the boundary of the dual of the component. For the example in Figure 5.3, the image of $f^\triangle$ restricted to the boundary cycles twice around the completely labelled vertex v_*, but in opposite direction: Following the boundary of the component in anti-clockwise direction in $X_*^\triangle$, the resulting paths runs in clockwise direction around v_*. Hence, the index of the component is -2. As in the case of the Index Lemma, the index counts, for a fixed $k \in I$, the number of almost completely labelled points with labels $I - \{k\}$ on the boundary of the dual of the component, where each such point is counted by is local orientation. For the example in Figure 5.3, there are two points on the boundary of the dual of the component with labels $1, 3$, both of which are oriented in the opposite way as the point with labels $1, 3$ on the boundary of $X^\triangle$. The same holds when considering points with labels $1, 2$ or $2, 3$.

As another example, consider the 3×4 game with an outside option for player II as shown below.

$$
\begin{bmatrix}
13, 13 & 7, 12 & 1, 14 & 0, 9 \\
12, 7 & 8, 8 & 2, 1 & 0, 9 \\
14, 1 & 1, 2 & 1, 1 & 0, 9
\end{bmatrix} . \tag{5.3}
$$

This is the game G^{+2} (1.15) as in Chapter 1. The outside option has, by the same reasoning as before, index $+2$. Figure 5.4 depicts the division of X into best reply regions and the dual construction $X_*^\triangle$ on the top. The dual of the component is depicted on the bottom. For the above example, the mapping $f^\triangle$ restricted to the boundary of the dual of the component yields a path running twice around v_*. This time, the orientations of the boundary and its image agree. For every $k \in I = \{1, 2, 3\}$, there are exactly two points on the boundary of the dual of the component with labels $I - \{k\}$ and whose orientation is the same as that of the point on the boundary of $X^\triangle$ with labels $I - \{k\}$. Therefore, the index of this component is $+2$.

These observations can be formalised as follows. Consider an $m \times n$ bimatrix game with an outside option for player II. Note that it is not necessary to assume that $m \leq n$. Let C denote the outside option equilibrium component.

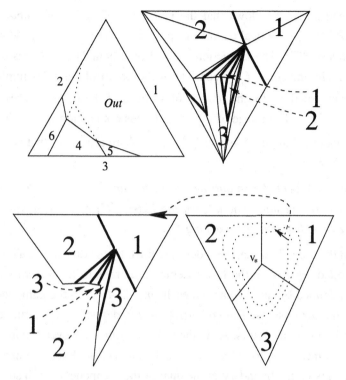

Fig. 5.4. An outside option component with index $+2$

Let V be the set of those vertices in player I's strategy space X that have *Out* as a best reply, so $V = \{v \in V \mid Out \in L(v)\}$. Now take the union of those $v^{\triangle}$ for which $v \in V$, so $C^{\triangle} = \bigcup_{v \in V} v^{\triangle}$. This union is referred to as the *dual of the component C* or the *dual of the outside option equilibrium component*. For generic outside options, the region $X(Out)$, i.e. the region in X where *Out* is a best reply, is a full-dimensional and convex region with vertices that have m labels (or it is empty). Hence, the set $C^{\triangle}$ is a union of $(m-1)$-simplices. These simplices yield a triangulation of $C^{\triangle}$. If v_{Out} denotes the vertex in $C^{\triangle}$ that represents the best reply region with label *Out*, then $C^{\triangle}$ is star-shaped with respect to v_{Out}. This follows from the fact that $C^{\triangle}$ is a union of simplices who all have v_{Out} as a vertex.

The boundary of $C^{\triangle}$ is denoted as $\partial C^{\triangle}$. The simplex $v^{\triangle}$ is an $(m-1)$-simplex for all $v \in V$, and the boundary $\partial C^{\triangle}$ is the union of the $(m-2)$-faces in $C^{\triangle}$ that do not include the vertex that represents *Out*. From the dual construction it follows that the pair $\left(C^\triangle, \partial C^\triangle\right)$ is homeomorphic to $\left(\triangle^{m-1}, \partial\triangle^{m-1}\right)$. The dual payoff mapping $f^\triangle$ as in (3.4) is well-defined on the boundary $\partial C^\triangle$. The restriction of $f^\triangle$ to the boundary of $C^\triangle$ is denoted as $f^\triangle_{|\partial C^\triangle}$. The image of $f^\triangle_{|\partial C^\triangle}$ consists of the union of $(m-2)$-simplices in $X^\triangle$ that are spanned by the images of vertices of the $(m-2)$-faces on the boundary of $C^\triangle$. The image of $f^\triangle_{|\partial C^\triangle}$ itself does not contain v_*. So the image of $f^\triangle_{|\partial C^\triangle}$ can be thought of as some $(m-2)$-sphere around v_* that consists of $(m-2)$-faces.

Definition 5.3. *Let C be an outside option equilibrium component of a game with a generic outside option. Then the index $I(C)$ of the component C is defined as the degree of the mapping $f^\triangle_{|\partial C^\triangle}$.*

So, as in the Index Lemma, the index is defined by the division of a boundary into labelled regions. In the Index Lemma, the regions arise from the mapping f^S, defined by unit vectors on each $(m-2)$-face. In the game theoretic context, the regions arise from the mapping $f^\triangle$, defined by a mixture of payoff vectors and unit vectors. As in the Index Lemma, however, the index of a component measures, for a fixed label k, the number of almost completely labelled points on the boundary of the dual of the component. Each such point is counted with its local orientation, and the measure does not depend on the choice of k.

Note that the image of $f^\triangle_{|\partial C^\triangle}$ can be retracted to the boundary of $X^\triangle$. This works in the same way as Lemma 4.5: If p is a point in the image of $f^\triangle_{|\partial C^\triangle}$, define the retraction as the intersection of the line between v_* and p, in the direction of p, with the boundary of $X^\triangle$. Note that v_* does not lie in the image of $f^\triangle_{|\partial C^\triangle}$. This is due to the non-degeneracy of the payoffs representing other strategies than *Out*.

For generic outside options, only payoff perturbations for player I in the outside option are of relevance. This can also be seen using the labelled dual construction. Small perturbations of the payoff matrix B leave the combinatorial structure of $|X^\triangle|$ invariant, since the combinatorial structure of the best reply regions in X is unaffected. Small perturbations of the payoff matrix A leave the combinatorial division of $\partial C^\triangle$ into best reply regions invariant, since for all simplices $v^\triangle$ and their faces that do not involve *Out*, the combinatorial division into best reply regions is invariant with respect to small perturbations. It follows from Definition 5.3 that small perturbations of the payoffs leave the index $I(C)$ invariant. Perturbations of player I's payoffs in the outside option, however, split $C^\triangle$ generically into labelled regions and determine those points in the interior of $C^\triangle$ that are mapped to v_*. These are the Nash equilibria that "survive" perturbations of the payoffs.

The local degree of $f^\triangle$ at these pre-images is the index of the equilibrium (see Lemma 3.15). But the sum of local degrees equals the degree of the mapping, which is again the same as the degree of $f^\triangle$ restricted to the boundary of the dual of the component. As a consequence, one obtains the following, well-known result.

Corollary 5.4. *Let the index of a generic outside option equilibrium component be $I(C)$. Then every small generic perturbation yields equilibria close to the component C such that the indices of these equilibria add up to $I(C)$.*

Proof. The proof follows the same lines as the proof of the Index Lemma, and is a consequence of the fact that the degree of a mapping is the same as the degree of a mapping restricted to its boundary.

An illustration of the proof is given in Figure 5.5 for a perturbation of G^{-2} as in (1.17) (compare Figure 5.3). The perturbation that is depicted is given by the payoff vector $(\varepsilon, 0, 0)^\top$ for player I in the outside option. For the illustration, ε is chosen to be large. It should be noted, however, that the combinatorial division of the dual of the component does not depend on the magnitude of ε (see also Lemma 6.4 in Chapter 6).

The combinatorial and geometric properties of the mapping $f^\triangle_{|\partial C^\triangle}$ are not affected by small perturbations. Generic perturbations, however, perturb the dual payoff mapping $f^\triangle$ in the interior of $C^\triangle$. Let the restriction of $f^\triangle$ to $C^\triangle$ be denoted as $f^\triangle_{|C^\triangle}$. Thus every small generic perturbation of the game gives a mapping $f^\triangle_{|C^\triangle} : C^\triangle \longrightarrow X^\triangle$. Although the mapping itself does depend on the perturbation, the index $I(C)$ does not, since the degree of $f^\triangle_{|\partial C^\triangle}$ stays invariant under small perturbations for the reasons explained above. The payoff perturbation renders the game generic and, hence, yields a generic division of $C^\triangle$ into labelled best reply regions (see Figure 5.5).

The degree of $f^\triangle_{|C^\triangle}$ is the same as the degree of $f^\triangle_{|\partial C^\triangle}$, and can be computed as the sum of local degrees at the pre-images of v_* in $C^\triangle$. These are the completely labelled points in $C^\triangle$ that represent equilibria in which *Out* is

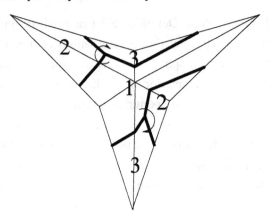

Fig. 5.5. A perturbation of an index -2 component

played with positive probability. This local degree is the same as the index of an equilibrium.

Since the perturbation is generic, these pre-images lie in the interior of some $v^\triangle$ in $C^\triangle$ and, for small perturbations, lie close to the vertex that represents *Out*.   □

For example, in Figure 5.5 one obtains two completely labelled points that read $1, 2, 3$ in clockwise direction, i.e. both have index -1. As noted above, Figure 5.5 depicts the case for a large ε. For a small ε, the completely labelled points lie close to the original vertex representing *Out*, but the combinatorial division stays invariant.

Corollary 5.4 is of course not a new result (see Section 1.3). New, however, is how it relates to the Index Lemma. In the Index Lemma, the index was defined as the degree of f^S on the boundary. For outside options it is the degree of $f^\triangle$ on the boundary of the dual of the component. Although f^S arises from unit vectors while $f^\triangle$ arises from general payoff vectors, in both cases the division of the boundary into labelled regions determines the sum of orientations of completely labelled points (or simplices) in the interior. As for the Index Lemma, one can summarise the result under 'The index equals the content". The boundary condition (i.e. the degree of the mapping on the boundary of the dual of the component) determines the number of completely labelled points in the interior of the dual of the component (i.e. the Nash equilibria that use *Out*), accounting for orientation.

5.3 Degenerate Games and General Equilibrium Components

This section describes how the dual construction might be applied to other components of equilibria. For example, the above analysis does not require that the payoffs for player II in the component are constant and independent of player I's strategy choice (as it is the case for outside options). Nevertheless, there are limits to the application of the dual construction to general components of equilibria in degenerate bimatrix games.

Take an $m \times n$ bimatrix game. If the payoffs for player II are non-degenerate, the triangulation $|X^\triangle|$ is well-defined. Furthermore, the dual payoff mapping $f^\triangle$ in (3.4) is well-defined since the payoff mapping f is well-defined. It is easy to verify that the Nash equilibria correspond with those points that are mapped to v_* under $f^\triangle$. So the Nash equilibria still correspond to completely labelled points. This follows from the definition of the payoff mapping f as in (3.3) via the artificial payoff matrix. The difference is that completely labelled points might, for example, lie on the boundary of a simplex $v^\triangle$, or that almost completely labelled points lie on some lower dimensional k-face of some $v^\triangle$ for $k < m - 2$. Also, there can be connected sets of completely labelled points in the labelled dual construction.

The latter case is illustrated by the following example.

$$\begin{bmatrix} 0,0 & 10,10 & 0,0 & 0,-10 \\ 0,0 & 0,0 & 0,10 & 0,8 \\ 0,10 & 0,0 & 10,0 & 0,8 \end{bmatrix} \tag{5.4}$$

This is a variant of Example 2.3. Against strategies 4 and 7 of player II, player I is indifferent between strategies $1, 2$ and 3. So the equilibrium component here is for player I to play some strategy in the union of the best reply regions $X(4)$ and $X(7)$, and for player II to play a best reply strategy, which is either strategy 4 or 7, or a mixture of both. In the latter case, the strategy of player I lies in the intersection of the best reply regions $X(4)$ and $X(7)$, and player II can play any mixture between strategies 4 and 7.

The dual of this component is depicted in Figure 5.6, in which the union of the best reply regions $X(4)$ and $X(7)$ is represented by a dashed line between the vertices that represent the best reply regions with labels 4 and 7. The mapping $f^\triangle$ is well-defined. In particular, it is well-defined on the boundary of the dual of the component C, and has degree zero: There is no point on the boundary of the dual of the component with labels $2,3$, and there are exactly two points on the boundary with labels $1,2$, and exactly two points with labels $1,3$. Each such pair of points is such that one almost completely labelled point has the opposite orientation of the other almost completely labelled point.

Hence, every (small) perturbation that makes the payoffs of player I generic yields a game with equilibria involving strategies 4 or 7 and whose indices add up to zero. Take, for example, the original game as in Example 2.3. This game is a perturbation of player I's payoffs in strategies 4 and 7, and has two equilibria using strategies with labels 4 or 7 and whose indices add up to zero. Multiplying the columns of A representing strategies 4 and 7 with some small constant $\varepsilon > 0$ yields a game with the same combinatorial properties that is close to the original game (see also Lemma 6.4).

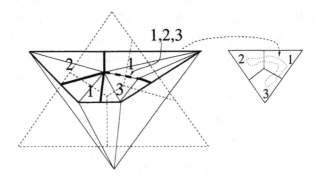

Fig. 5.6. The dual of the component in (5.4)

The problem is that, in general, degeneracies occur in the payoff matrices of both players. Furthermore, components (and hence their duals) are not necessarily homeomorphic to some simplex. This limits the direct application of the dual construction to general components of equilibria. Consider, for example, the following game constructed by Kohlberg and Mertens (1986):

$$
\begin{bmatrix}
1,1 & 0,-1 & -1,1 \\
-1,0 & 0,0 & -1,0 \\
1,-1 & 0,-1 & -2,-2
\end{bmatrix}
\tag{5.5}
$$

In this example, the equilibrium component is a cycle, both in player I's as well as in player II's strategy space. It can easily be verified that the component in (5.5) has index $+1$. It is the unique component, and strategies 1 and 4 weakly dominate the other strategies, so a slight perturbation only leaves one pure strategy equilibrium. The dual construction cannot be applied directly, since neither the "vertices" in X nor the "vertices" in Y are well-defined, i.e. they have more than three labels. For example, the "vertex" corresponding to pure strategy 1 by player I has labels $2, 3$ (the unplayed strategies) and $4, 6$ (best replies). Thus neither $X^\triangle$ nor $Y^\triangle$ are well-defined.

Nevertheless, there are ways of still applying the dual construction to such components. Take an $m \times n$ bimatrix game (with $m \leq n$). Then the payoffs for, say, player II, can be made non-degenerate by small payoff perturbations. Then $|X^\triangle|$ is well-defined for the perturbed payoff matrix B. This then yields the mapping $f^\triangle$ and a division of $|X^\triangle|$ into labelled regions. The drawback of this approach is that the dual construction $|X^\triangle|$ and hence $f^\triangle$ are not independent of the payoff perturbations used for player II.

6

Index Zero and Hyperstability

This chapter shows that outside option equilibrium components that have index zero are not hyperessential. This yields a characterisation of hyperessentiality of outside option equilibrium components in terms of the index: An outside option equilibrium component is hyperessential if and only if it has non-zero index. In a parallel and independent work, Govindan and Wilson (2004) show that the result presented here for outside option equilibrium components also holds for general equilibrium components in N-player games. The merit of the approach presented here is that it requires only basic tools from algebraic topology and provides a geometric intuition.

An equilibrium component is said to be *essential* if for every small perturbation of the game there exists an equilibrium of the perturbed game that is close to the component (Wu and Jiang (1962); Jiang (1963)). Kohlberg and Mertens (1986) extend the concept of essentiality to perturbations of all equivalent games, i.e. games obtained by adding convex combinations of existing strategies as pure strategies. A component is referred to as *hyperessential* if it is essential in all equivalent games. They define a component that is a minimal hyperessential component as *hyperstable*.

This chapter addresses the question how (hyper)essentiality in a game theoretic context and essentiality in a topological context (i.e. non-zero index) are linked (see e.g. Govindan and Wilson (1997a;b) for a discussion). It is a well-established fact that topological essentiality implies strategic essentiality. The converse, however, is not true, as an example of an equilibrium component with index zero that is essential shows (Hauk and Hurkens

(2003)). However, until recently, it was unknown whether hyperessentiality implies topological essentiality. This question is answered affirmatively for outside option equilibrium components in bimatrix games by employing the dual construction to outside option components.

The structure of this chapter is as follows. Given the similarities between the Index Lemma and the index for outside option equilibrium components, Section 6.1 studies index zero labellings in case of the Index Lemma. It is shown that for every index zero boundary labelling there exists a triangulation and a labelling (subject to the division on the boundary) such that the triangulation does not contain a completely labelled simplex (Theorem 6.1). Section 6.2 reviews the concepts of essentiality and hyperessentiality, and it is shown how the results for index zero labellings apply to index zero outside option equilibrium components. It is shown that an outside option equilibrium component is hyperessential if and only if it has non-zero index (Theorem 6.7). The result is based on duplicating the outside option, which yields a refinement of the triangulation of the dual of the component. This allows one to divide the dual of the component into labelled regions such that no point is completely labelled. This work concludes with Section 6.3. It gives an example of an outside option equilibrium component that is essential in all equivalent games that do not contain a copy of the outside option (Lemma 6.10).

6.1 Index Zero Labellings

This section discusses index zero labellings for triangulations of $(m-1)$-dimensional polytopes P. Given a triangulation of $|\partial P|$ into $(m-2)$-simplices with a labelling L of the vertices of $|\partial P|$, the definition of the index as in Definition 5.1 is well-defined via the Sperner mapping f^S. The Index Lemma implies that every labelled triangulation of P that agrees with the given triangulation and labelling on ∂P must contain completely labelled simplices whose orientations add up to the index of the labelling on the boundary. This section shows that if the boundary labelling on ∂P has index zero, then there exists a labelled triangulation of P that agrees with the given triangulation and labelling on ∂P and that does not contain a completely labelled simplex.

Let P be an $(m-1)$-dimensional polytope. Furthermore, let $|\partial P|$ be a triangulation of ∂P into $(m-2)$-simplices together with a labelling of the vertices of $|\partial P|$. This defines the Sperner mapping f^S on the boundary ∂P as in (3.1). The index of the boundary labelling is defined as the degree of f^S restricted to the boundary and counts, for a given label $k \in I$, the almost completely labelled points on the boundary with labels $I - \{k\}$, accounting for their orientation. The following results for labellings as in the Index Lemma might not be new (Theorems 6.1 and 6.3). The author, however, is not aware of results as stated below in the literature.

Theorem 6.1. *Let $|\partial P|$ be a labelled triangulation of ∂P into $(m-2)$-simplices with index zero. Then there exists a labelled triangulation $|P|$ that agrees with the given labelled triangulation of the boundary and that does not contain a completely labelled simplex.*

Proof. Let $f_{|\partial P}$ denote the restriction of f^S to the boundary. The fact that $\deg f^S_{|\partial P} = 0$ implies that $f^S_{|\partial P}$ is homotopic to some constant map via a homotopy h (see e.g. Bredon (1994, II, Corollary 16.5 and V, Lemma 11.13)). This means that $f^S \simeq_h \star$, where $\star$ denotes some constant map. In other words, there exists a mapping $h : \partial P \times [0,1] \rightarrow \partial \triangle_\star^{m-1}$ such that $h(x,0) = f^S(x)$ and $h(x,1) = \star$ for all $x \in \partial P$. Since h is constant on $\partial P \times 1$, one obtains a mapping, which is also denoted as h, from $\partial P \times [0,1]_{/\sim(\cdot,1)}$ to $\partial \triangle_\star^{m-1}$, where $\partial P \times [0,1]_{/\sim(\cdot,1)}$ denotes the quotient space that is generated by the equivalence relation that identifies $(\cdot, 1)$ with a single point; the space $\partial P \times [0,1]_{/\sim(\cdot,1)}$ can be thought of as a "cone" over ∂P, which is homeomorphic to P.

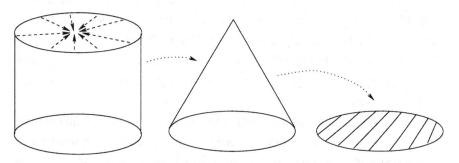

Fig. 6.1. The cone over ∂P

This is depicted in Figure 6.1 for P being the 2-dimensional disk. The boundary of the disk is the 1-dimensional sphere S^1. Then $S^1 \times [0, 1]$ is a cylinder as depicted on the left. Identifying $(\cdot, 1)$ with a single point yields the "cone" as depicted in the middle, which is homeomorphic to the 2-dimensional disk depicted on the right.

Thus h can be seen as a mapping $h : P \longrightarrow \partial \triangle_*^{m-1}$ that agrees with f^S on the boundary. This is a well-known result that states that a mapping from the unit $(m-1)$-sphere to the unit $(m-1)$-sphere that has degree zero can be extended to a mapping from the unit m-ball D^m to the unit $(m-1)$-sphere. The result goes back to Hopf (see e.g. Bredon (1994) as cited above).

The mapping h divides P into labelled regions which are the pre-images of the regions in $\triangle_*^{m-1}$. This is depicted in Figure 6.2. Now choose a triangulation of P with no vertices on the boundary other than the original vertices on ∂P. This can, if necessary, be achieved by adding a single vertex in the centre of P, since P is convex. Next, choose an iterated pseudo refinement of this triangulation that allows vertices on the boundary and that is such that each simplex is smaller in diameter than some given $\delta > 0$. Now label every vertex in the interior of $|P|$ according to $L(v) \in L(h(v))$, where $L(h(v))$ are the labels of the image of v in $\triangle_*^{m-1}$ (see Figure 6.2). There is no point on the boundary $\partial \triangle_*^{m-1}$ that has all m labels, so no simplex in the refinement can have more than $m - 1$ distinct labels, as long as the simplices are sufficiently small. Notice that, since P is compact, the mapping h is uniformly continuous.

Finally, one has to get rid of the vertices that were added to the boundary ∂P. This works in the same way as in Lemma 4.4, since the labelling of vertices on the boundary is *consistent*. That is, if a vertex v lies on an k-face of the original triangulation spanned by original vertices $v_1, \dots, v_k$, then $L(v) \in \{L(v_1), \dots, L(v_k)\}$. This is the labelling equivalent to the consistency as in Definition 4.3.

So let the vertices that were added by the iterated pseudo refinement be $v_1, \dots, v_n$, and let Λ be the ordered index set of the vertices that were added to the boundary. Let v be a vertex on the boundary. Then $v = \sum_{i=1}^{l} \mu_i v_i$ with $\mu_i > 0$, for some $v_1, \dots, v_l$. In particular, the labelling satisfies $L(v) = L(v_i)$ for some $i \in \{1, \dots, l\}$. So the face spanned by $\{v_1, \dots, v_{i-1}, v, v_{i+1}, \dots, v_k\}$ has

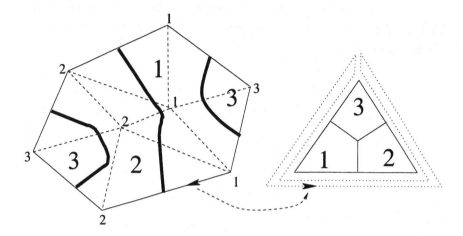

Fig. 6.2. A labelling with index zero

the same labels as the face spanned by $\{v_1, \ldots, v_{i-1}, v_i, v_{i+1}, \ldots, v_k\}$. A simplex spanned by $\{v_1, \ldots, v_{i-1}, v, v_{i+1}, \ldots, v_k\}$ and some $\{v_{k+1}, \ldots, v_m\}$ is fully labelled if and only if the simplex spanned by $\{v_1, \ldots, v_{i-1}, v_i, v_{i+1}, \ldots, v_k\}$ and $\{v_{k+1}, \ldots, v_m\}$ is fully labelled.

So the vertices that were added by the iterated pseudo refinement and that lie on the boundary of ∂P can be removed (in reverse order) to obtain a refinement with no vertices added to the boundary and no completely labelled simplex. □

Remark 6.2. In Figure 6.2, the Sperner mapping f^S on the boundary has index zero, but is onto. Suppose one is restricted in subdividing P. For example, assume a triangulation $|P|$ with the same boundary labelling as in Figure 6.2, but that has only one vertex in the interior of P. This is depicted in Figure 6.3. Then every labelling of the interior vertex yields (pairs of) completely labelled simplices. The reason is that the interior vertex is connected to all boundary faces. For every label $k \in \{1, 2, 3\}$, there are faces on the boundary with missing label k, that is, faces with labels $1, 2$ or $2, 3$ or $1, 3$. These almost completely labelled faces come in pairs of opposite orientation because of the index zero property. Thus, in the restricted case, one always obtains completely labelled simplices whose orientations add up to zero. In the next section, it is shown how this restricted case compares with the essentiality of an equilibrium component as in the example by Hauk and Hurkens (2002), and how the unrestricted case compares with the hyperessentiality of an equilibrium component.

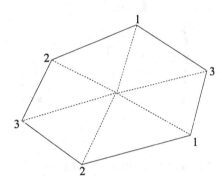

Fig. 6.3. A labelling with index zero and a restricted triangulation

For non-zero labellings one obtains the following result.

Theorem 6.3. *Let* $|\partial P|$ *be a labelled triangulation of* ∂P *with index k. Then there exists a labelled triangulation* $|P|$ *that agrees with the given labelled triangulation of the boundary and is such that* $|P|$ *contains* $|k|$ *completely labelled simplices, each with orientation* sign *k.*

Proof. The idea is to divide P into labelled regions such that there exist exactly $|k|$ completely labelled points in P with orientation sign k. This division is then covered by small simplices.

Choose a subset B in the interior of P that is homeomorphic to an $(m-1)$-ball. Define a mapping $f_{|\partial B}$ on the boundary of B that maps the boundary of B on $\partial \triangle_*^{m-1}$ and that is such that each almost completely labelled point on the boundary of $\triangle_*^{m-1}$ has exactly $|k|$ pre-images in ∂B with orientation sign k. Such a mapping exists and can be constructed as follows. Identify the boundary ∂B with the unit sphere S^{m-1}. For $(x_1, \cdots, x_m) \in S^{m-1}$, the tuple (x_1, x_2) can be seen as a complex number z, and the mapping $f_{|\partial B}(z, x_3, \cdots, x_m) = (z^k, x_3, \cdots, x_m)$ will do.

The mapping $f_{|\partial B}$ has the same degree as the Sperner mapping f^S on the boundary of P. Hence, the mapping f^S restricted to the boundary ∂P and $f_{|\partial B}$

are homotopic via some homotopy, denoted as h. The homotopy h can be identified with a mapping from $P - B$ to $\partial \triangle_*^{m-1}$, since $[0, 1] \times \partial P$ is homeomorphic to $P - B$. Note that ∂B and ∂P are homeomorphic to $\partial \triangle_*^{m-1}$, and are hence themselves homeomorphic. This yields a division $P - B$ into labelled regions with no completely labelled point. Label the region B with some arbitrary but fixed label. Then the division of P into labelled regions is such that there exist exactly $|k|$ points that are completely labelled. These lie on the boundary of B. This is depicted in Figure 6.4 for a boundary mapping with index $+1$.

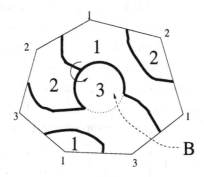

Fig. 6.4. Obtaining a division with exactly $|k|$ completely labelled points

From here, the proof follows the same lines as the proof of Theorem 6.1. Cover P with sufficiently small simplices and label the vertices according to the regions they are contained in. The vertices that are added to the boundary of P can be omitted by the same argument as in the proof of Theorem 6.1 and Lemma 4.4.   □

As explained in Chapter 5, there are strong similarities between the situation in the Index Lemma and outside option equilibrium components. The next section shows how the results from above translate into the game theoretic context and how one can divide the dual of an outside option into best reply regions, given the boundary division, such that it does not contain a completely labelled point, i.e. an equilibrium. This can be achieved by duplicating the outside option only.

6.2 Index Zero Outside Option Equilibrium Components

In this section, it is shown that an outside option equilibrium component (in a bimatrix game with generic outside option) is hyperessential if and only if it has non zero index. It is also explained how the results of the previous section fit in the game theoretic context. Before proving the main result of this section, the concepts of essentiality and hyperessentiality are briefly reviewed.

Wu and Jiang (1962) define essential Nash equilibria. The extension to compact sets of Nash equilibria is described by Jiang (1963), and is also discussed in van Damme (1991, Section 10.2). In analogy to the concept of essential fixed point sets (Fort (1950)), an equilibrium component C of a game G is called *essential* if and only if for every small payoff perturbation of the game G there exists an equilibrium of the perturbed game that is close to C. A game $\tilde{G}$ is called an equivalent game to G if $\tilde{G}$ can be obtained from G by adding a finite number of convex combinations of strategies of G as pure strategies. In other words, the games G and $\tilde{G}$ have the same reduced normal form. For example, the two games shown below are equivalent.

$$G = \begin{bmatrix} 10,10 & 0,0 \\ 0,0 & 10,10 \end{bmatrix} ; \quad \tilde{G} = \begin{bmatrix} 10,10 & 0,0 & 5,5 & 3,3 \\ 0,0 & 10,10 & 5,5 & 7,7 \\ 1,1 & 9,9 & 5,5 & \frac{33}{5},\frac{33}{5} \end{bmatrix}$$

A strategy in an equivalent game can be interpreted as a strategy of the original game and vice versa by rescaling the probabilities for the strategies. An equilibrium component C of a game G is referred to as *hyperessential* if it is essential in all equivalent games $\tilde{G}$. Kohlberg and Mertens (1986) define a set S as *hyperstable* if it is minimal with respect to the following property: S is a closed set of Nash equilibria of G such that, for any equivalent game, and for every perturbation of the normal form of that game, there is a Nash equilibrium close to S. It follows that a hyperessential equilibrium component must contain a hyperstable set (Kohlberg and Mertens (1986)): Let F denote the family of subsets of a single connected component that is hyperessential, ordered by set inclusion. Every decreasing chain of elements in F has a lower bound, and therefore, applying Zorn's Lemma, the family F must have a minimal element.

It is a well-established fact that non zero equilibrium components are both essential and hyperessential. The index of a Nash equilibrium component is invariant under addition or deletion of redundant strategies Govindan and Wilson (1997a, Theorem 2; 2004, Theorem A.3). Therefore the index of a component is the same in all equivalent games. Since the index measures the sum of indices of equilibria close to the component if the game is slightly perturbed, a non-zero index implies both essentiality and hyperessentiality of the component (see also Section 1.3 for the properties of the index).

Whether the converse is also true was an open question until recently. In fixed point theory, a component of fixed points under a mapping f is called essential if every mapping close to f has fixed points close to the component. O'Neill (1953) shows that a fixed point component is essential if an only if it has non-zero index. In game theory, the Nash equilibria can be described as the fixed points of a map. A perturbation of the game yields a mapping for the game that is close to the original fixed point mapping.

So the question arises whether, by suitably perturbing the game, one can show equivalence between strategic and topological essentiality. Referring to the results of O'Neill (1954), Govindan and Wilson (1997b) write: "The resolution of this puzzle is important for axiomatic studies because in a decision-theoretic development it would be implausible to impose topological essentiality as an axiom unless it is provable that the space of games is rich enough to obtain equivalence between strategic and topological essentiality."

Hauk and Hurkens (2003) found an example of a bimatrix game with an outside option in which the outside option equilibrium component has index zero and that is nonetheless essential. This shows that game theoretic and topological essentiality are not equivalent. If restricted to perturbations of the original game, the space of games is not rich enough to obtain equivalence between topological and strategic essentiality. However, their example fails the requirement of hyperessentiality. So the question arises whether the concept of hyperessentiality is the game theoretic equivalent of topological essentiality.

In this section, it is shown that this is the case for outside option equilibrium components with a generic outside option. Furthermore, it is demonstrated why an index-zero component can be strategically essential, but not hyperessential. Comparing it with the case of the Index Lemma, essentiality compares with a triangulation in which one is restricted in the number of

simplices in the subdivision, and hyperessentiality compares with the unre-
stricted case (see Remark 6.2). Govindan and Wilson (2004), in a parallel and
independent work, show that index zero components cannot be hyperessential
in general. Their approach is discussed at the end of this section. The merit
of the proof presented here is that it only needs basic tools from algebraic
topology. Also, since the dual construction can easily be visualised, it also
provides a geometric and combinatorial intuition for the result.

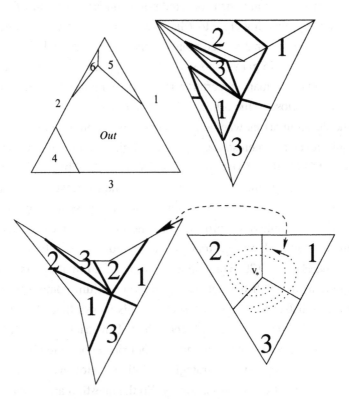

Fig. 6.5. An index zero essential component

The idea of the proof can be explained by considering an example of an
outside option equilibrium component that is essential but not hyperessential.
Such an example is given by the game in (6.1). This is the game by Hauk and
Hurkens (2002) showing that topological essentiality is not the equivalent of
topological essentiality.

$$\begin{bmatrix} 4,5 & 0,-23 & 2,-1 & 0,0 \\ 0,-15 & 8,-1 & -2,-21 & 0,0 \\ 2,-11 & 1,3 & 3,1 & 0,0 \end{bmatrix} \tag{6.1}$$

The dual construction for this game is given in Figure 6.5. The dual payoff mapping $f^{\triangle}$, restricted to the boundary of the dual of the outside option component, has degree zero. The image does not complete a full cycle. Hence, the outside option equilibrium component has index zero. This can also be verified by a simple counting argument. There is only one other equilibrium of the game, namely the pure strategy equilibrium with payoffs $(4,5)$.

Hauk and Hurkens show that the component is essential. It should be noted that only payoff perturbations of the payoffs for player I in the outside option are of importance. All other payoffs are generic. Looking at the dual construction of the game, it can be seen that the restricted dual payoff mapping $f^{\triangle}_{|\partial C^{\triangle}} : \partial C^{\triangle} \to X^{\triangle}$ is such that the image of $f^{\triangle}_{|\partial C^{\triangle}}$ "wraps" completely around v_*, but does not complete a full cycle.

A more detailed depiction of the image of $f^{\triangle}_{|\partial C^{\triangle}}$ is given in Figure 6.6. The image of $f^{\triangle}_{|\partial C^{\triangle}}$ consists of a union of $(m-2)$-simplices in $X^{\triangle}$. These are the images of the faces of $C^{\triangle}$, and are depicted in bold dashed lines. In the figure, v_{Out} is the image under $f^{\triangle}$ of the vertex in $X^{\triangle}$ that represents best reply region Out in X, and the vertices v_l are the images of the vertices in $X^{\triangle}$ that represent a best reply region with label l or an unplayed strategy l in X $(l = 2,5,6)$.

Now suppose one perturbs the payoffs in the outside option. Then v_{Out} lies close to v_*. Consider, for example, a perturbation of Out such that strategy 1 of player I is the best reply to Out. Then v_{Out} lies in the region with label 1 close to v_*, as depicted in Figure 6.6. So there are two simplices in the image of $C^{\triangle}$ that contain v_*, namely the simplex spanned by v_5, v_6 and v_{Out} and the simplex spanned by v_6, v_2 and v_{Out}. The former simplex represents the vertex in X with labels 5, 6 and Out, the latter represents the vertex in X with labels 6, unplayed strategy 2 and Out. A similar analysis applies if v_{Out} lies in one of the regions with label 2 or 3. Therefore, the component is essential. This is the game theoretic counterpart to the situation described in Remark 6.2.

It should be noted, however, that it is not sufficient to just count the almost completely labelled points on the boundary of a component to see whether a

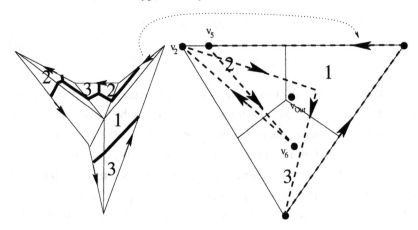

Fig. 6.6. The essentiality of the component

component is essential or not. The payoff mapping is generally more complex than the Sperner mapping, since the payoff vectors are generally not unit vectors. Consider, for example, the component depicted in Figure 6.7. This component is similar to that of game (6.1). The difference is that the payoffs for player I in the column of (6.1) representing strategy 6 are modified such that v_6 is shifted to the left compared with v_6 in Figure 6.6. There are two points on the boundary of $C^\triangle$ with labels $1, 2$, two with labels $1, 3$ and two with labels $2, 3$, and each pair is such that the points have opposite orientation. But the component is not essential. There is a "gap" in the image around v_*. If the perturbation of the outside option for player I were such that v_{Out} lies in the shaded area as depicted, then there would not exist an equilibrium that uses *Out*. A necessary and sufficient condition for the essentiality of a component is that the retraction of the image of $\partial C^\triangle$ is onto. The retraction is defined as on page 110 for components and is similar to that described in Lemma 4.5: If p is a point in the image of $f^\triangle_{|\partial C^\triangle}$, define the retraction as the intersection of the line between v_* and p, in the direction of p, with the boundary $X^\triangle$. This condition ensures that there is no "gap" in the image of $\partial C^\triangle$, so the image "wraps" completely around v_*.

Now suppose one duplicates *Out* and perturbs the payoff for player II such that the original regions in X where *Out* is a best reply is divided as depicted in Figure 6.8. This yields two vertices in the dual construction that are associated with the outside option. Hence, by looking at equivalent games in which

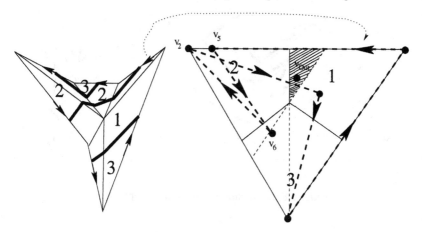

Fig. 6.7. A non-essential component

Out is duplicated, one obtains "richer" divisions of $C^\triangle$ into best reply regions. For example, if one makes strategy 2 of player I the best reply to Out_1, and strategy 1 the best reply to Out_2, one obtains a perturbation of the equivalent game that has no equilibrium close to the component. The associated labelled dual of this perturbed equivalent game is illustrated in Figure 6.8. Since there is no completely labelled point in the dual of the outside option, there is no equilibrium that involves *Out*, and hence no equilibrium close to it. The associated payoff perturbations are given in (6.2).

$$\begin{bmatrix} 4,5 & 0,-23 & 2,-1 & 0,0 & \varepsilon,0 \\ 0,-15 & 8,-1 & -2,-21 & \varepsilon,0 & 0,\varepsilon \\ 2,-11 & 1,3 & 3,1 & 0,2\varepsilon & 0,0 \end{bmatrix} \tag{6.2}$$

The method of duplicating *Out* is the underlying idea in the proof of Theorem 6.7. The idea is to divide the dual of the component into labelled regions such that there exists no completely labelled point, as in Theorem 6.1. One then has to show that such a division can in fact be created by duplicating *Out* and perturbing the payoffs in the duplicates of *Out*. Duplicating *Out* and perturbing the payoffs for player II in the duplicates refines the triangulation of $C^\triangle$ into simplices $v^\triangle$. The difference to Theorem 6.1 is that the new vertices are close to the vertex representing *Out*. Perturbing the payoffs for player I then divides the simplices in the refined triangulation into labelled regions.

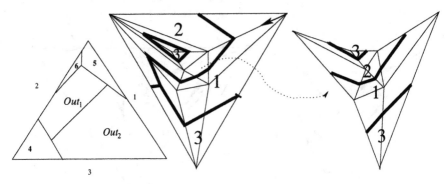

Fig. 6.8. Duplication of the outside option

Unlike the proof of Theorem 6.1, this is achieved by assigning payoffs to the vertices, as opposed to assigning labels.

Consider an outside option game with a generic outside option for player II. It is first shown that the magnitude of the perturbations for player I in the outside option does not matter when analysing the essentiality of an outside option equilibrium component. The following lemma shows first that the combinatorial division of $X_*^{\triangle}$ into simplices and labelled regions is invariant under multiplying payoff columns of player I with some positive constant. Two $m \times n$ games are referred to as *combinatorially equivalent* if both yield combinatorially equivalent triangulations $|X^{\triangle}|$ and if the divisions of the simplices in the triangulation are combinatorially the same.

Lemma 6.4. *Let G be an $m \times n$ bimatrix game represented by payoff matrices A and B. Let $\tilde{G}$ be represented by $\tilde{A} = [\lambda_1 A_1, \ldots, \lambda_n A_n]$ and B, where $\lambda_j > 0$, for $j = 1 \ldots, n$. Then G and $\tilde{G}$ are combinatorially equivalent.*

Proof. Let $\lambda_1 > 0$ and $\lambda_j = 0$ for $j \neq 1$. Let (x, y) be a Nash equilibrium of G. Define $\tilde{y}' = (\frac{y_1}{\lambda_1}, y_2, \ldots, y_n)$. Rescaling $\tilde{y}'$ such that it lies in Y yields $\tilde{y}$ such that $(x, \tilde{y})$ is a Nash equilibrium of $\tilde{G}$. Continuing in the same fashion with the other λ_j yields the desired result. $\square$

Lemma 6.4 shows that the combinatorial equilibrium properties of a game are unaffected if a column of A or a row of B is multiplied by some positive constant. One just has to adjust the weights on the strategies to account for the multiplication of the columns and rows. It also shows that the combinatorial structure of $|X^\triangle|$ and the combinatorial division $X^\triangle_*$ is invariant under such operations. As a corollary one obtains the following result.

Corollary 6.5. *Let G be a game with outside option for player II in which the outside option equilibrium component has index zero. Let $\tilde{G}$ be obtained from G by copying Out a finite number of times. If there exists a perturbation of $\tilde{G}$ with small payoff perturbations for player II and large payoff perturbations for player I in the copies of Out such that there is no equilibrium that plays a copy of Out with positive probability, then there exists a small perturbation of $\tilde{G}$ such that there exists no equilibrium close to the outside option equilibrium component.*

Proof. Without loss of generality it can be assumed that the payoffs to player I in the outside option are zero. Adding or subtracting some constant to the payoff columns of A does not change the best reply properties. The payoffs for player I in $\tilde{G}$ can be described as follows.

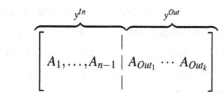

Let (y^{In}, y^{Out}) be a strategy profile that makes player I indifferent between best reply strategies $i_1, \ldots, i_k$. Now multiply the columns A_{Out_j} by some $\varepsilon > 0$, and consider the strategy $(\frac{y^{In}}{c}, \frac{y^{Out}/\varepsilon}{c})$, where $c = \sum_j y^{In}_j + \sum_l \frac{y^{Out}_l}{\varepsilon}$. Then strategies $i_1, \ldots, i_k$ are still the best reply strategies. Thus one can easily switch from large perturbations to small perturbations for player I in copies of *Out*, and vice versa, without changing the equilibrium properties of the game. $\square$

The proof of Theorem 6.7 below uses a similar argument as in Corollary 6.5 for the payoff perturbations for player II in the copies of *Out*. In the proof of Theorem 6.7 one divides the dual of an outside option into smaller simplices by adding vertices. These vertices correspond to added strategies for player II. The following lemma shows that one can obtain a combinatorially equivalent refinement such that the added vertices are close to the vertex representing *Out*. Any two vertices that are close have payoffs to player II that are close. This follows from Lemma 2.2. Two triangulations with vertices $v_{k \in K}$ and $v'_{k \in K}$ are called *combinatorially equivalent* if the affine linear extension of $g(v_k) = v'_k$, $k \in K$, on the vertices is an isomorphism that maps simplices on simplices and faces on faces.

Lemma 6.6. *Let $C^\triangle$ be the dual of an outside option equilibrium component, and let v_{Out} denote the vertex in $C^\triangle$ representing Out. Consider an iterated refinement of $C^\triangle$ with no vertices added to the boundary of $C^\triangle$. Then there exists a combinatorially equivalent iterated refinement in which the added vertices are close to v_{Out}.*

Proof. The proof is by induction on the number of added vertices. Note that $C^\triangle$ is star-shaped (see page 109). So the case is clear for just one added vertex.

Now suppose one has an iterated refinement with k added vertices. Consider the refinement that is obtained by adding the first $k - 1$ vertices. For this refinement, there exists a combinatorially equivalent refinement with $k - 1$ vertices close to v_{Out}. The vertex added last in the iterated refinement lies in some simplex in this refinement (which might not be unique, in case it lies on some face). This simplex corresponds to a simplex in the refinement where all vertices are close to v_{Out}. Hence, one can add a vertex close to v_{Out} to the $k - 1$ other vertices close to v_{Out} in order to obtain a combinatorially equivalent iterated refinement. □

The following theorem is the game theoretic equivalent of Theorem 6.1. The index is given by a division of the boundary into labelled regions. If the index is zero, this division can be extended to a division of $C^\triangle$ such that no point in $C^\triangle$ is completely labelled. As in the proof of Theorem 4.6, one then has to account for the restriction imposed by the game theoretic context. In particular, one has to show that this division can be achieved by perturbing an equivalent game in which *Out* is duplicated a finite number of times.

Theorem 6.7. *Let C be an outside option equilibrium component in a generic outside option game. Then C is hyperessential if and only if $I(C) \neq 0$.*

Proof. Without loss of generality assume that all payoffs for player I are positive and that the payoffs in the columns of A add up to 1, i.e. $|A_j| = 1$ (this can be achieved by first adding a suitable constant to each column and then scaling; see Section 3.3). Let $I(C) = 0$, so the dual payoff mapping $f^\triangle_{|\partial C^\triangle}$ has degree zero. Instead of considering the dual payoff mapping $f^\triangle_{|\partial C^\triangle}$, it is more convenient to consider the payoff mapping f and its restriction $f_{|\partial C^\triangle}$ to the boundary $\partial C^\triangle$. Note that $f^\triangle$ is simply $\mathrm{Id}^\triangle \circ f$. In particular, the image of $f^\triangle_{|\partial C^\triangle}$ completes a cycle around v_* if and only if the image of $f_{|\partial C^\triangle}$ completes a cycle around v_*. Therefore, the mapping $f_{|\partial C^\triangle}$ has also degree 0.

It follows that $f_{|\partial C^\triangle}$ is homotopic to some constant map $\star$ (see e.g. Bredon (1994, II, Corollary 16.5 and V, Lemma 11.13)), where the constant lies on the boundary of $\triangle_*^{m-1}$. First the mapping can be retracted to the boundary of $\triangle_*^{m-1}$ (see Lemma 4.5 and p. 110), and can then be deformed into a constant map along $\triangle_*^{m-1}$. Let this homotopy be denoted as h. So $h: \partial C^\triangle \times [0,1] \to \triangle_*^{m-1}$, and v_* does not lie in the image of h.

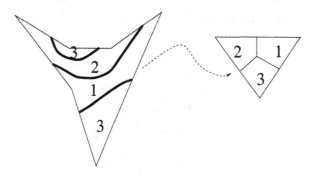

Fig. 6.9. A homotopy for outside option equilibrium components

As in the proof of Theorem 6.1, the mapping $f_{|\partial C^\triangle}$ extends to a mapping on $C^\triangle$ such that no point is mapped on v_*. This can be seen as follows. The homotopy is constant on $(\partial C^\triangle, 1)$. This yields $h: (\partial C^\triangle \times [0,1])_{/\simeq(\cdot,1)} \to \triangle_*^{m-1}$, where $\partial C^\triangle \times 1$ is identified with a single point. The dual component $C^\triangle$ is star-shaped (see page 109), so $(\partial C^\triangle \times [0,1])_{/\simeq(\cdot,1)}$ is homeomorphic to $C^\triangle$. This gives a mapping, also denoted as h, that maps $C^\triangle \to \triangle_*^{m-1}$ such that v_* does not lie in the image of h. The pre-images of the labelled regions in $\triangle_*^{m-1}$ now divide $C^\triangle$ into labelled regions such that no point in $C^\triangle$ is completely labelled. This is depicted in Figure 6.9 for the component in the example (6.1).

One now has to show that such a division can be achieved in a game theoretic context as a division into best reply regions by refining the triangulation of $C^\triangle$ and choosing the payoffs for player I accordingly. For this, as in the proof of Theorem 4.6, choose an iterated pseudo refinement of the triangulation of $C^\triangle$ that allows one to add vertices to the boundary of $C^\triangle$. Now assign a payoff $h(v)$ to each vertex v in the iterated pseudo refinement. Then the payoffs $h(v)$ for vertices added to the boundary are consistent with the payoffs for the original vertices on the boundary of $C^\triangle$. If the simplices in the refinement have a sufficiently small diameter, the image of a simplex is a simplex in $\triangle_*^{m-1}$ that does not contain v_*. This is ensured by h being uniformly continuous.

Now delete all vertices that were added to the boundary of $|C^\triangle|$. According to Lemma 4.4, this does not create completely labelled points, and, by Lemma 4.2, yields a regular triangulation. This results in a division of $C^\triangle$ as depicted in Figure 6.10 for the component in the example (6.1).

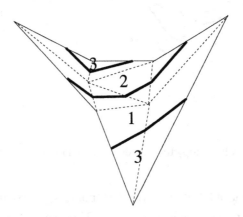

Fig. 6.10. An approximation of the homotopy

So far, one has created an extended game in which strategies for player II are added (see Lemma 3.12). Each added vertex corresponds to an added strategy. The corresponding payoffs to player II in the added strategies are determined by Lemma 2.2, and those for player I are given by the value of the homotopy at the vertex that represents the added strategy. The extended game is such that neither *Out* nor any of the added strategies are played in an equilibrium.

It remains to show that a similar game, i.e. one that yields a combinatorially equivalent division of $C^\triangle$ into simplices and best reply regions, can be created as a perturbed equivalent game. This is achieved by duplicating *Out* and perturbing the payoffs in the copies of *Out*.

Let $v_{k \in K}$ be the set of vertices added, where K is an ordered set, reflecting the order in which the vertices were added. From the above construction each vertex v_k has a payoff $h(v_k)$. Lemma 6.6 shows that there exists a combinatorially equivalent refinement of $C^\triangle$ in which all added vertices lie close to v_{Out}, the vertex representing *Out* in $C^\triangle$. Let the set of the vertices in this refinement be denoted as $v'_{k \in K}$, where v'_k is close to v_{Out} and corresponds to v_k.

Now assign the payoffs $h(v_k)$ to vertex v'_k. This yields a division of $C^\triangle$ into best reply regions that is combinatorially equivalent to the original division. In particular, it does not contain a completely labelled point. This is depicted in Figure 6.11 for the component in (6.1).

Now every vertex in $|X^\triangle|$ that is close to the vertex v_{Out} has payoffs to player II that are close to the payoffs of *Out* to player II if the regular triangulation is translated into an extended payoff matrix B' (see Lemma 2.2). So B' consists of B and perturbed copies of *Out*. As for the payoffs $h(v'_k)$ for player I, Corollary 6.5 shows that one can make them arbitrarily small without creating equilibria. Hence, one created a game that is a perturbed equivalent game in which the outside option is duplicated a finite number of times.   □

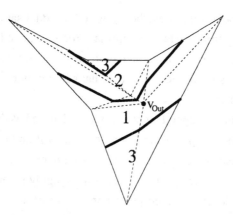

Fig. 6.11. Adding vertices close to v_{Out}

In the same way as an outside option equilibrium component with index zero might be essential (i.e. having at least $2l$ ($l > 0$) equilibria for every small perturbation), an index k outside option equilibrium component might have $|k| + 2l$ ($l > 0$) equilibria for every small perturbation of the original game. Using the dual construction, such an example would be easy to create (a $3 \times n$ game would be sufficient for that). Allowing perturbations of equivalent games, one gets, similarly to Theorem 6.3, the following result.

Proposition 6.8. *Let C be an outside option equilibrium component with index $I(C) = k$. Then there exists an equivalent game and a perturbation of the equivalent game such that there are only $|k|$ equilibria close to C and whose indices add up to k.*

Proof. The proof follows the same lines as the one of Theorem 6.7, and is the game theoretic equivalent of Theorem 6.3. If the index of a component is $I(C) = k$, then there exists a homotopy between the payoff mapping $f_{|\partial C^\triangle}$ and a mapping that maps an $(m-2)$-ball exactly k times around itself. This homotopy is used to divide $C^\triangle$ into labelled regions such that there exist exactly $|k|$ completely labelled points in $C^\triangle$ with local degree sign k (as in the proof of Theorem 6.3). Then this division of $C^\triangle$ can be imitated by duplicating *Out* a sufficient number of times and choosing the payoffs for player I accordingly, just as in the proof of Theorem 6.7. □

Section 5.3 above discusses the limits of the dualisation methods with respect to general components of equilibria. Problems arise from the fact that, in general, degeneracies occur in the payoff space of both players. Therefore, the above method is insufficient to prove that general index zero components cannot be hyperessential.

In a parallel and independent work, Govindan and Wilson (2004) show that an equilibrium component has non-zero index if and only if it is hyperessential. Their results are based on results from fixed point theory and apply to general N-player games, and their proof uses highly technical arguments.

In fixed point theory, a fixed point component of a mapping f is called essential if every mapping close to f has fixed points close to the component (Fort (1950)). It is a well-known result in fixed point theory that if the fixed point index of a component is zero, and if the underlying space is "well behaved", then there exists a fixed point free mapping close to the original mapping (O'Neill (1953)). In game theory, the Nash equilibria can be described as the fixed points of a suitable mapping. A perturbation of the game yields a mapping for the perturbed game that is close to the original fixed point mapping. The Hauk and Hurkens example and the example presented in the next section, however, show that just considering perturbations of the original game is not sufficient to obtain equivalence between strategic and topological essentiality.

The index of a component is the same in all equivalent games (Govindan and Wilson (1997a, Theorem 2; 2004, Theorem A.3)). By considering equivalent games, one increases the space of possible perturbations. Thus the space of mappings that can be obtained from perturbing equivalent games increases in dimension. This is the underlying idea in the proof of Govindan and Wilson for general components of equilibria. The authors show that, if allowing equivalent games, the space of games, i.e. the space of perturbed equivalent games, is rich enough to obtain equivalence between topological and game theoretic essentiality.

The authors start from a map that has no fixed points close to the component. Such a map exists after O'Neill (1953). From this map the authors create a perturbed equivalent game that is such that the Nash map for this game, i.e. the mapping that describes the Nash equilibria of the game as fixed points, copies the properties of the original fixed point free map. That is, the Nash map does not have fixed points close to the component. Thus a component is hyperessential if and only if it has non-zero index.

In essence, the key idea of the approach by Govindan and Wilson and of the approach presented here is the same. One has the existence of mappings with certain properties. For outside option components, the mapping does not map a point in the dual of the component to the completely labelled point. Considering the parallels with the Index Lemma, the index reflects a combinatorial property of the component. In the case of Govindan and Wilson, one has a fixed point free mapping. The index describes a topological property of the component. By adding redundant strategies it is shown that the these mappings can arise as mappings from a perturbed equivalent game.

Remark 6.9. The combinatorial nature of the approach presented above is such that, by duplicating *Out*, one creates one equivalent game such that, for all $\varepsilon > 0$, there exists a perturbation of that game smaller than ε that has no nearby equilibria. In particular, the equivalent game is independent of ε. This is not the case for the equivalent game constructed by Govindan and Wilson (2004), where the equivalent game depends on ε. Typically, one has to add more and more redundant strategies as ε becomes smaller.

6.3 Restricted Duplication of Strategies and Index Zero: An Example

Hauk and Hurkens (2002) show the non-hyperessentiality of the component in the game (6.1) by adding a convex combination of strategies as a new strategy for player I, i.e. not by duplicating *Out*. The added strategy is a convex combination of strategies 1 and 2 (for details see Hauk and Hurkens (2002)).

This section provides an example of an index zero outside option equilibrium component that is not only essential, but is essential in all equivalent games that do not contain a duplicate of *Out*. It shows that duplicating *Out* is not only sufficient, but in cases also necessary to create an equivalent game in which an index zero outside option equilibrium component is not essential. For general index zero equilibrium components, this suggests that it is necessary to add redundant strategies for both players in order to create an equivalent game in which the component is not essential.

The example is constructed as follows. Consider the following game.

$$G^0 = \begin{bmatrix} H^2 & 0 & 0,9 \\ 0 & H^- & 0,9 \end{bmatrix}, \tag{6.3}$$

with

$$H^2 = \begin{bmatrix} 10,10 & 0,0 \\ 0,0 & 10,10 \end{bmatrix}, \qquad H^- = \begin{bmatrix} 13,13 & 7,12 & 1,14 \\ 12,7 & 8,8 & 2,1 \\ 14,1 & 1,2 & 1,1 \end{bmatrix}. \tag{6.4}$$

Game G^0 is the same as the game in (1.18) in Section 1.4. The 2×2 game H^2 in the upper left part in G^0 is a 2×2 coordination game, and the 3×3 game H^- in the lower middle part of G^0 is a game where the mixed strategy equilibrium in which both players mix uniformly between their first two strategies yields the highest equilibrium payoff, which is 10 to both players (see also (1.13) and (1.16) for further discussion). In Section 1.4, it is shown that the outside option equilibrium component of the game G^0 has index 0. The only equilibria that are not "cut off" by the outside option are the pure strategy equilibria in H^2 and the mixed strategy equilibrium in H^- with payoff 10 for both players. The two former ones have index +1, the latter one has index -1. Hence, the outside option equilibrium component has index 0.

Lemma 6.10. *The outside option equilibrium component $C(G^0)$ of the game in (6.3) is essential in all equivalent games that do not contain a duplicate of Out. In particular, the component is essential.*

Proof. Consider the games $\tilde{G}^2$ and $\tilde{G}^{-1}$ as below.

$$\tilde{G}^{-1} = \begin{bmatrix} H^{+2} & \overset{0,9}{\vdots} \\ 0,0 & {}_{0,9} \end{bmatrix}, \qquad \tilde{G}^2 = \begin{bmatrix} H^- & \overset{0,9}{\vdots} \\ 0,0 & {}_{0,9} \end{bmatrix} \qquad (6.5)$$

Then the outside option equilibrium components in $\tilde{G}^2$ and $\tilde{G}^{-1}$ are both essential and hyperessential. The games $\tilde{G}^2$ and $\tilde{G}^{-1}$ are variants of the games G^2 as in (1.15) and G^{-1} as in (1.17). By the same reasoning as in Section 1.4, it is easy to verify that $C(\tilde{G}^2)$ has index +2, and that $C(\tilde{G}^{-1})$ has index -1, where $C(\cdot)$ denotes the outside option equilibrium component of a game. Thus both $C(\tilde{G}^2)$ and $C(\tilde{G}^{-1})$ are essential and hyperessential. Now consider the equivalent game, denoted as $\tilde{G}^0$, in which one adds convex combinations for player I. Then every such game is of the form

$$\tilde{G}^0 = \begin{bmatrix} H^{+2} & 0,0 & 0,9 \\ \hline \nleq 9 \leq 9 & \vdots \\ \hline \leq 9 \leq 9 & \vdots \\ \hline \leq 9 \nleq 9 & 0,9 \\ \hline 0,0 & H^- & \vdots \end{bmatrix}, \qquad (6.6)$$

where the entry '$\nleq 9$' means that at least one payoff for player II in that part of the game is larger than 9, and '≤ 9' means that all the payoffs for player II in that part of the matrix are less than or equal to 9. Note that the payoffs in H^{+2} and H^- are such that a convex combination does not allow entries larger than 9 in both parts of a row, i.e. in both the H^{+2} and the H^- part of a convex combination of original columns. It is now sufficient to consider only payoff perturbations for player I in the outside option, since all other payoffs of the game $\tilde{G}^0$ are generic. Let the perturbation vectors of player I's payoffs in the outside option be denoted by ε^u, ε^m and ε^l for perturbations in the upper, middle and lower part of the game (6.6). Without loss of generality it can be assumed that $\varepsilon^u \geq 0$, $\varepsilon^m \geq 0$ and $\varepsilon^l \geq 0$. It can also be assumed that the perturbation is generic, i.e. there is a unique maximal perturbation. Suppose there were two (or more) maximal perturbations. If one is among the ε_i^u and one among the ε_i^l, then player I mixing uniformly between the strategies with the maximal perturbation and player II playing *Out* is an equilibrium close $C(\tilde{G}^0)$. All other cases of non-generic perturbations are covered by the three cases below.

1) The maximal perturbation is among the ε_i^m. In this case, player I playing the strategy with that maximal perturbation and player II playing *Out* is an equilibrium close to $C(\tilde{G}^0)$.
2) The maximal perturbation is among the ε_i^u. Then consider the game consisting of the first two strategies of player II and *Out* and the strategies as in (6.6) for player I, with payoffs and perturbations as above, i.e. consider

$$T = \begin{bmatrix} H+2 & \varepsilon_1^u,9 & \vdots \\ \not\leq 9 & \vdots \\ \hline \leq 9 & \varepsilon_1^m,9 \\ & \vdots \\ \hline \leq 9 & \varepsilon_1^l,9 \\ & \vdots \\ 0,0 & \vdots \end{bmatrix}, \tag{6.7}$$

T is an perturbed equivalent form of the game $\tilde{G}^{-1}$ in (6.5). Since $C(\tilde{G}^{-1})$ is hyperessential, there exists a strategy pair (x,y) that is an equilibrium close to the outside option equilibrium component $C(\tilde{G}^{-1})$. It is now shown that this strategy pair, if interpreted as a strategy pair of the game $\tilde{G}^0$, is also an equilibrium close to $C(G^0)$. First consider player I. By construction, player I has no incentive to deviate from the strategy x, seen as a strategy of the game $\tilde{G}^0$ as in (6.6), if player II plays strategy y as a strategy of the game $\tilde{G}^0$.

It remains to show that player II has no incentive to deviate from y, seen as a strategy for the game $\tilde{G}^0$ via the mapping $(y_1, y_2, y_{Out}) \mapsto (y_1, y_2, 0, 0, 0, y_{Out})$. The strategy profile x is such that the first two strategies of player II must yield a payoff of less than or equal to 9, where at least one must yield a payoff of 9. Otherwise, player II would play *Out* only, and this cannot be an equilibrium for the game T due to the maximal perturbation ε_i^l. But, by the choice of the payoffs in the games H^{+2} and H^-, this means that the other strategies of player II's (except for *Out*) cannot be best replies against x, i.e. they all yield a payoff strictly less than 9. This is because either the first strategy of player I or the second strategy of player I must have a weight of around $\frac{9}{10}$. This implies that the remaining weight is not sufficient to yield an expected payoff larger than 9 for player II in the other strategies (except from *Out*). Thus (x, y) is an equilibrium of the game $\tilde{G}^0$, which is also close to $C(G^0)$.

3) The maximal perturbation is among the ε_i^l. Then consider the game consisting of the third, fourth and fifth strategy of player II and *Out* and the strategies as in $\tilde{G}^0$ for player I, with payoffs and perturbations as above, i.e. consider

$$
T' = \begin{bmatrix} 0,0 & \overset{\varepsilon_1'',9}{\vdots} \\ \hline \leq 9 & \vdots \\ \hline \leq 9 & \overset{\varepsilon_1''',9}{\vdots} \\ \hline \not\leq 9 & \overset{\varepsilon_1',9}{\vdots} \\ \hline H^- & \vdots \end{bmatrix}, \tag{6.8}
$$

Then the analysis is analogous to the one above. The game T' is a perturbed equivalent form of the game $\tilde{G}^2$ in (6.5). The component $C(\tilde{G}^2)$ is both essential and hyperessential. Thus there exists an equilibrium (x, y) of T' that is close to $C(\tilde{G}^2)$. In the same way as above it can be verified that (x, y) is also an equilibrium of the game $\tilde{G}^0$ that is close to $C(G^0)$.

Thus the component is essential in all equivalent games of the form (6.6). It remains to show that it is also essential when adding convex combinations for player II, but no copies of *Out*. For this, extend the game T as in (6.7) by three columns of zeros, and the game T' as in (6.8) by two columns of zeros.

Then the index of the components in these modified games stays invariant, and the components remain hyperessential. Now consider the game $\tilde{G}^0$ as in (6.6) and add convex combinations of strategies for player II, but no duplicate of Out. If the maximal perturbation in the outside option lies in the upper part, the added convex combinations can be translated into convex combinations of the modified game T by assigning the weight on columns $3, 4, 5$ to the added columns of zeros in T. The component in the modified game T is hyperessential, and one shows that the equilibrium close to the component in the modified game T is also an equilibrium of the equivalent game of (6.6). For maximal perturbations in the lower part of the game one does the same analysis with the modified game S by treating the weights on columns $1, 2$ as weights on the two added columns of zeros. If the maximal perturbation lies in the middle part, the case is trivial. □

References

1. Bredon GE, (1993) Topology and Geometry. Graduate Texts in Mathematics. Springer, New York
2. Demichelis S, Germano F (2000) On the indices of zeros of Nash fields. Journal of Economic Theory 94: 192–217
3. Demichelis S, Ritzberger K (2003) From evolutionary to strategic stability. Journal of Economic Theory 113:51–75
4. Dold A (1972) Lectures on Algebraic Topology. Springer, New York
5. Eaves BC, Scarf H (1976) The solution of systems of piecewise linear equations. Mathematics of Operations Research 1:1–27
6. Fort MK (1950) Essential and non essential fixed points. Amer. J. Math. 72:315–322
7. Garcia CB, Zangwill WI (1981) Pathways to Solutions, Fixed Points and Equilibria. Prentice-Hall, Englewood Cliffs
8. Govindan S, von Schemde A, von Stengel B (2003) Symmetry and p-stability. International Journal of Game Theory 32:359–369
9. Govindan S, Wilson R (1997a) Equivalence and invariance of the index and degree of Nash equilibria. Games and Economic Behavior 21:56–61
10. Govindan S, Wilson R (1997b) Uniqueness of the index for Nash equilibria of two-player games. Economic Theory 10:541–549
11. Govindan S, Wilson R (2004) Characterization of hyperstability. Working paper
12. Gül F, Pearce D, Stacchetti E (1993) A bound on the proportion of pure strategy equilibria in generic games. Mathematics of Operations Research 18:548–552
13. Hauk E, Hurkens S (2002) On forward induction and evolutionary and strategic stability. Journal of Economic Theory 106:66–90

14. Henle M (1994) A Combinatorial Introduction to Topology. Dover Publications, New York

15. Hofbauer J (2000) Some thoughts on sustainable/learnable equilibria. Working paper

16. Jiang J-H (1963) Essential components of the set of fixed points of the multivalued mappings and its application to the theory of games. Scientia Sinica 12:951–964

17. Knaster B, Kuratowski C, Mazurkiewicz C (1929) Ein Beweis des Fixpunktsatzes für n-dimensionale Simplexe. Fundamenta Mathematicae 14:132–137

18. Kohlberg E, Mertens J-F (1986) On the strategic stability of equilibria. Econometrica 54:1003–1037

19. Lemke CE, Howson JT Jr (1964) Equilibrium points of bimatrix games. SIAM J. 12:413–423

20. McLennan A, Tourky R (2004) From Lemke-Howson to Kakutani. Working paper

21. Nash J (1951) Non-cooperative games. Annals of Mathematics 54:286–295

22. O'Neill B (1953) Essential sets and fixed points. Amer. J. Math. 58:497–509

23. Ritzberger K (2002) Foundations of Non-Cooperative Game Theory. Oxford University Press, Oxford

24. Savani R, von Stengel B (2004) Long Lemke-Howson paths. Working paper

25. Scarf HE (1983) Fixed point theorems and economic analysis. American Scientist 71:289–296

26. Shapley LS (1974) A note on the Lemke–Howson algorithm. Mathematical Programming Study 1: Pivoting and Extensions, 175–189

27. Spanier E (1966) Algebraic Topology. McGraw-Hill, New York

28. Sperner E (1928) Neuer Beweis für die Invarianz der Dimensionszahl und des Gebietes. Abh. Math. Sem. Univ. Hamburg 6:265–272

29. van Damme E (1989) Stable equilibria and forward induction. Journal of Economic Theory 48:476–496

30. van Damme E (1991) Stability and Perfection of Nash Equilibria. Springer-Verlag, Berlin

31. von Stengel B (1999a) Mixed strategy equilibria in bimatrix games. Lecture Notes in Game Theory, London School of Economics and Political Science, London

32. von Stengel B (1999b) New maximal numbers of equilibria in bimatrix games. Discrete and Computational Geometry 21:557–568

33. von Stengel B (2002) Computing equilibria for two-person games. In: Aumann RJ, Hart S (eds) Handbook of Game Theory, Vol. 3. North-Holland, Amsterdam
34. Wu W-T, Jiang J-H (1962) Essential equilibrium points of n-person non-cooperative games. Scientia Sinica 11:1307–1322
35. Ziegler GM (1995) Lectures on Polytopes. Springer, New York.

Symbolindex

Index

Lecture Notes in Economics and Mathematical Systems

For information about Vols. 1–464
please contact your bookseller or Springer-Verlag